# From Trotsky to Gödel

# From Trotsky to Gödel

## *The Life of Jean van Heijenoort*

Anita Burdman Feferman

A K Peters, Ltd.
Wellesley, Massachusetts

Editorial, Sales, and Customer Service Office

A K Peters, Ltd.
888 Worcester Street, Suite 230
Wellesley, MA 02482
www.akpeters.com



Cover designed by Hannus Design Associates.

Formerly published as
*Politics, Logic, and Love: The Life of Jean van Heijenoort*

**Library of Congress Cataloging-in-Publication Data**

Feferman, Anita Burdman.
[Politics, logic, and love]
From Trotsky to Gödel: the life of Jean van Heijenoort / Anita Burdman Feferman
p. cm.
"Formerly published as Politics, logic, and love : the life of Jean van Heijenoort."
Includes bibliographical references and index.
ISBN 1-56881-148-9 (alk. paper)
1. Van Heijenoort, Jean, 1912– 2. Trotsky, Leon, 1879–1940–Friends and associates.
3. Logicians–France–Biography. I. Title.

B2430.V36 F44 2000
194-dc21
[B]

00-065011

Printed in the United States of America
10 09 08 07 06 10 9 8 7 6 5 4

*To Helen Grand Feferman,*
*who also remembers*

The author is extremely grateful to the following people for their invaluable contributions of photographs and help in obtaining illustrations for this book:

Paulette Corpion and Jean van Heijenoort *fils,* for photographs from the van Heijenoort family collection (front jacket cover, Chapters One, Two, Seven, Eight, Twelve, Thirteen);

Esteban Volkov and the Trotsky Institute in Coyoacán, Mexico, Albert Glotzer, and David King, for photographs from their personal and archival collections that deal with van Heijenoort's years with Trotsky (Chapters Four through Nine);

Stefan Bauer-Mengelberg and Marjorie Jonas, for their personal photographs in Chapter Thirteen;

Jean Gollay, for the photograph in Tenampa;

Solomon Feferman, for the photographs of Tlayacapán in Chapter Fifteen, of the Panthéon Français in Mexico City (Chapter Sixteen), and of the author (back jacket flap);

Jacques Faucher, for the photographs of van Heijenoort in Grenoble, including back jacket photograph.

Claude Imbert, for her perseverance in obtaining the illustration of the Lycée Saint-Louis from the Library of the History of Paris;

Pierre Movilliat, for providing the illustration of the Collège de Clermont from the archives of the Ville de Clermont de l'Oise;

Hélène Laroche Davis, for providing the photograph of Jean Marais.

The author is responsible for the photographs of van Heijenoort in the Fefermans' living room.

The author is grateful for permission to quote from previously published sources as follows:

Reprinted by permission of the publishers from *With Trotsky in Exile,* by Jean van Heijenoort. Cambridge, Mass: Harvard University Press, © 1978 by the President and Fellows of Harvard College.

Reprinted by permission of the Harvard Library Bulletin, "The History of Trotsky's Papers," by Jean van Heijenoort, July 1980.

# Contents

III



IV

# Acknowledgments

THIS BOOK GREW out of what I had thought would be an interview or two with Jean van Heijenoort. That was like thinking I could eat two peanuts; once I got started it was almost impossible to stop. Along the way, the kindness, generosity, and interest of van Heijenoort's family, friends, colleagues, students, wives, and companions added immeasurably to my information and understanding. Now that my work is finished, it gives me great pleasure to be able to thank these good people. First, I am deeply indebted to Laure van Heijenoort and Jean van Heijenoort, *fils*, for sharing their memories, insights, and knowledge of their father, as well as for their important comments on a draft of the manuscript, and I must thank Paulette Corpion, van Heijenoort's half-sister, for her contributions of family history and lore. I am grateful beyond measure to Stefan Bauer-Mengelberg, Van's friend and colleague for forty years, who, besides contributing vividly detailed memories, also read and made useful suggestions on a draft version of the book.

In personal interviews, and also by telephone and written correspondence, I gained perspective as well as further indispensable information about my subject from many other kind people. I am indebted to Anne B., Dr. Lea Beaussier, Charles Boltenhouse, the late Margie Borisoff, Norman Borisoff, Felipé Bracho and Ana Luz Trejo-Lerdo, who were also my indispensable guides in Mexico, Pierre Broué, who read an early version of the manuscript and offered essential information about Trotsky's inner circle, Douglas Bryant, Joel Carmichael, Jacques Cohen, Manuel Corpion, Giovanna Corsi, Elena Danielson, Martin Davis, Philippe de Rouilhan, Rodney Dennis, Burton Dreben, Pascal Engel, Anne Fagot-Largeault, Olivia Gall, Jean-Yves Girard, Warren Goldfarb, the late Benjamin Gollay, Jean Gollay, Patricia Griscom, Loretta B. Guyer, the late Sidney Hook, Claude Imbert, Edith Jeffrey, Richard Jeffrey, Marjorie Jonas, Marlene Kadar, Peter Katel, Joseph Keller, Minna Klotz, Georg Kreisel, Peter Lax, Ruth Marcus, Pierre Movilliat, Sidney Morgenbesser, Pablo Noriega, Salvador Nuños Traslosheros, Daniella Montelatici Prawitz, Wilfried Sieg, Joseph Stillman, Frederick Stoker, Ezequiel Valdés Avila, Patricia Valdés Vidal, Zenaida Vidal de Valdés, Carine van Heijenoort, Yveline van Heijenoort, Françoise Ville, Esteban Volkov, Palmira Volkov, William Walker, Judson Webb, Beverly Woodward, and the late Ana María Zamora. In my research at Hoover Institution, I was given valuable assistance by Elena Danielson and Linda Wheeler.

For careful reading of all or part of my manuscript and useful comments, besides those already mentioned I would like to thank Catherine Barnett, Cheryl Dawson, John Dawson, Priscilla Feigen, Timothy Fernando, Gail Francis, Susan Gay, David Malament, Carol Parikh, Rohit Parikh, and Constance Reid. I must also give special thanks to Mary Lowenthal Felstiner, Joseph Frank, Diane Wood Middlebrook, and my colleagues in the biographers' seminar at Stanford University. The value of the intellectual stimulation and encouragement offered by this group of people of wide-ranging interests was enormous.

Initially, Francesco del Franco of Naples encouraged me to write this biography rather than a shorter work, and arranged to publish it with his fine press, Bibliopolis. He later graciously consented to its publication by a larger house, thus permitting wider distribution. My publisher, Klaus Peters of Jones and Bartlett, provided constant encouragement and support, and Lucy Ferriss, who edited the manuscript, added to the grace and clarity of the text. I want to thank the staff at Jones and Bartlett: in particular, Natasha Sabath for her excellent work on the manuscript, and Heather Stratton.

Finally, there is always the person a book could not have been written without. For me, Solomon Feferman, patient, kind, loving, understanding, and possessed of every other virtue under the sun, including a willingness to read and discuss, is the one. He knows.

ANITA BURDMAN FEFERMAN
*Stanford, California*

# Introduction

SECRETIVE IN LIFE, enigmatic in death, Jean van Heijenoort was a man of many mysteries. To begin with, his name, pronounced van Hi'-en-ort, was Dutch, but he was born in France and was French to the core. His family called him Jean, most of his friends called him Van, and in the United States, for many years, he signed himself John. Futhermore, a French dictionary of biography lists thirteen aliases for Jean van Heijenoort: Alex Barbon, Cestero, Garcia Cestero, Jarvis Gerland, Marcel Letourneur, Daniel Logan, Marc Loris, Marcel, Karl Meyer, Jean Rebel, Jean Vannier, Ann Vincent, and J. Walter Wind.

When he was killed, in 1986, at the age of seventy-three, van Heijenoort was internationally known and respected as a logician and a historian of modern logic. His last academic base was Stanford University where he had been working as an editor on the *Collected Works of Kurt Gödel*; before that he had been a professor at Brandeis and New York University. Van Heijenoort also had a second, less widely-known career as a consultant and expert in special acquisitions for Harvard library's vast Trotsky Archive. But, not surprisingly, the news reports of his death focused upon a third career, one he had abandoned forty years earlier. From 1932 to 1945 van Heijenoort had been a full-time, professional revolutionary, and for the first seven of those years he had been Leon Trotsky's personal secretary, bodyguard, and translator, living in exile and wandering the globe with his leader and idol. The newspaper headlines told an even more sensational story: van Heijenoort had been murdered in Mexico City by his fourth wife, who then committed suicide. In almost every account of the tragedy, the eerie echo of Trotsky's assassination (in Mexico, forty-six years earlier) could be heard, but it was not political passion that killed van Heijenoort.

Unlike most biographers, I met my subject when he was alive, and had the good fortune to see him often. We ate dinners together, went to concerts, movies, and attended the same parties. I heard his voice and his laugh, and took note of his quick gestures and his weightless walk. There was something shadowy, withdrawn, yet alluring in his manner that drew me to him. He played his cards close to the chest, but occasionally he would show his hand. After I had been interviewing him for two years he began to say, only half in jest, "You know me better than I know myself." I felt privileged and, at the same time, aware that I was approaching an edge, and that it was time to draw back.

I first met him in 1970 in Paris. I knew almost nothing about him then except that he was a logician, like my husband. Tall, angular, with bony, chiseled features, van Heijenoort was an attractive man; he was reserved, even uncomfortable in company, but not unfriendly, and he had a sly wit. On the surface, he was the dignified professor deeply involved in his field of symbolic logic, which in its abstractness is about as far from politics as you can get. But smoldering beneath that subdued, restrained manner was a look, an unmistakable intensity, that begged to be discovered. Immediately, I wanted to ask questions, but the signs said Do Not Touch.

Two things stuck in my mind from that first meeting. After dinner with a small group, we had walked through the Latin Quarter. Pointing to a building on the corner of the Boulevard Saint-Michel and Rue Vaugirard, van Heijenoort said, "That's the prison where I was locked up for two years." The "prison" turned out to be his old school, the Lycée Saint-Louis. Emboldened by this unexpected opening in the conversation, which until that point had been carefully polite, I blurted out, "Aha, so you are not Dutch."

He replied, "My father was Dutch, but I never knew him."

Then, as if I had flicked a switch, the light left his face and it was clear that the subject was closed. Not until 1983, when I began work on this book, would I learn the details about van Heijenoort's absent father, as well as what he had meant about being locked up in the "lycée-prison," and that his escape had coincided with his leap into revolutionary politics. By that time, I had heard a few rumors about his having once been Trotsky's secretary and bodyguard, but since he seemed so unlikely a candidate, I had assumed it must have been for a very short period, for if ever a man did not fit the stereotype of the wild-eyed radical revolutionary, it was this pallid, introverted scholar.

Even after he had written *With Trotsky in Exile* (Harvard University Press, 1978), a memoir of his seven years as a member of Leon Trotsky's intimate circle, van Heijenoort seldom spoke of his life as Comrade Van, and when he did, it was only to those he thought were ready to understand. In public, he never was tempted to begin a sentence with, "Well, you know, when I was with Trotsky ..." Was he still a Trotskyite?* Was he a liberal? A neo-conservative? Was he a Democrat or a Republican? His long-time friends and colleagues did not really know. The most he would offer was: "I am no longer in politics. I am apolitical." The one thing he did make clear was his contempt for those who merely dabbled in politics. For him, it was an all or nothing affair.

Almost everything Van did was an all or nothing affair. This was true of his private life, too, about which he was so circumspect that hardly anyone knew where he lived or with whom. To those who were privy to some of the details of his personal entanglements and four or five marriages ("depending upon how you count," Van said) he was the quintessential romantic Frenchman, a nomad who made his rounds between France, Mexico, and the United States, who loved women and forgot about logic and reason, as time after time he found himself involved in affairs and marriages which caused pain, remorse, and eventually led to his own violent end.

Although I saw Jean van Heijenoort on several occasions after our first meeting in Paris, I did not learn anything more about him or feel that I knew him better. However, after he came to Stanford in 1982 to begin work on the *Collected Works of Kurt Gödel* with my husband Solomon Feferman and others,

---

*In using the term 'Trotskyite' rather than 'Trotskyist', I have followed Jean van Heijenoort's lead. He always referred to himself and his comrades as Trotskyites, when writing or speaking (in English) of his political past. It has been called to my attention that some consider the '-ite' suffix pejorative in this connection. I do not, nor did Van.

he was frequently in my sight, often at my own dinner table. I was still intimidated by what I took to be his remoteness, but there were small cracks in the armor. He was polite, he was correct, he was tight-lipped, but a good meal and lively company loosened him up. My curiosity grew. Why and how did a man who labeled himself "timid, terribly shy," join Trotsky's inner circle as the Old Man's secretary and bodyguard? What was it like, living that life for all those years? In 1983, half expecting to be rebuffed outright or put off, I asked if I might interview him for an hour or two with an eye to writing a short profile. To my surprise, his response was an immediate, unhesitating yes, almost as if he had said, I thought you'd never ask. In spite of his modesty, something in his attitude acknowledged that he knew he was a worthy subject.

The "hour or two" turned into three years of interviews that eventually included not only him but members of his family, his wives, his friends, and his colleagues and students. Among other places, I visited Creil, where he was born; the village in Norway where Trotsky lived from 1935 to 1936 and where Van was arrested by the Norwegian police; Mexico, where he lived with Trotsky in the famous blue house (now the Frida Kahlo Museum) and became close friends with Frida Kahlo and Diego Rivera and André Breton; and East Hampton, Long Island, where Van built a little house with his own hands, and where Jackson Pollock, Willem de Kooning, Harold Rosenberg, and Dwight Macdonald were included in his circle of friends and neighbors.

Our first interview took place on a Saturday afternoon in van Heijenoort's office on the second floor of the mathematics building at Stanford University. It was a small room, neatly arranged and rather bare as academic offices go; most of his books were still at Harvard, in his Pusey Library office. Wearing loose blue jeans and a blue sweater, he looked youthful, not at all like a man in his seventieth year. He was friendly but wary. I worried that he might object to a tape recorder. On the contrary, to my relief, he regarded it as a useful tool and wanted to know the make, model, and what I thought of its quality.

When I began the interview by asking some general questions about his background, van Heijenoort stiffened in his chair and clasped his hands tightly. His first answers were formal. After a while, I asked what he remembered about his childhood, and, in particular, about his earliest memories. Then his position changed; he let go of his fingers; his shoulders relaxed, and he leaned forward with something important to tell me. As much as anything else, his astonishing answer was what moved me to write his story.

# I

## Chapter One

# In My Beginning Is My End

### Creil: 1912–1918

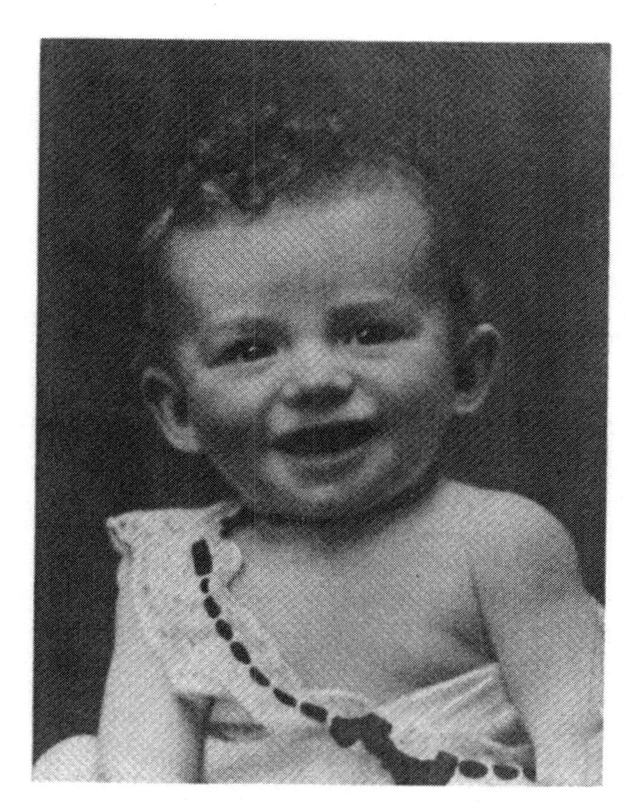

*Jean Louis Maxime van Heijenoort, 1913.*

I saw my father bleed to death. My mother was holding a big white basin and my father was spitting blood. I don't really remember my father, but I remember that. It was September 1914; the war had just begun; I was two years and two months. That was my arrival in this world: watching my father bleed to death.

JEAN VAN HEIJENOORT, *1983*

FEW PEOPLE have as clear a notion as Jean van Heijenoort did of the forces and events that have determined the course of their lives. There was no doubt in his mind that his father's death coupled with the outbreak of World War I defined the terms of his existence. The vivid and terrifying image of his father's death was the dominating motif of his life. As an adult he insisted that although he was only two at the time, the memory was *his,* and not the memory of a retelling. War and death were the heralds of his "arrival in this world." He meant, of course, not his physical birth, but the birth of his consciousness.

Jean Louis Maxime van Heijenoort was born on July 23, 1912, in Creil, a small town on the Oise River, fifty kilometers north of Paris. His mother, Charlotte Hélène Balagny and her ancestors had lived in that region for as long as anyone remembered, but his father, Jean Théodore Didier van Heijenoort, was an immigrant, a Dutchman who had come to France to seek his fortune.

Baby Jean was the first boy in the Balagny family in three generations. The singularity of his maleness in a family of women was impressed upon him early, and for the rest of his life it remained a contrapuntal theme he played against that of his father's death. As an adult, on the rare occasions when he spoke of his background, he would say, "I was raised in a milieu totally of women. I had my mother, but no father; my grandmother, but no grandfather; my aunt, but no uncle, and my two girl cousins who were much older than me." With raised eyebrows, open palms, and a self-mocking quizzical smile, he would add, "What *that* did to me, I don't know."

There was at least one positive result: Jean was at ease and easily affectionate among women, and he carried a lifelong, self-fulfilling expectation that this affection would be reciprocated. Not surprisingly, his first romance was with his mother, whom he adored, and his first painful jealousy was on her account. A small, dark, vivacious woman, she was the template against which he measured all his other loves. As Stefan Bauer-Mengelberg, one of his oldest friends, noted, "She ran through his life like a red thread."

Charlotte Hélène Balagny, always called Hélène, was born in 1887 to a family of very modest means. In van Heijenoort's words, his ancestors were "peasant stock or handicrafts people with little or no formal education." Hélène's mother had never attended school and could neither read nor write. Hélène, on the other hand, did go to elementary school and excelled as a student. Indeed her reputation for brilliance both in and out of school was such that members of her family, including van Heijenoort, later asserted that if she had had the proper education she would have become another Marie Curie.

The claim is, of course, impossible to substantiate, but it does illuminate all that was lacking in Hélène's education and environment. Curie, born twenty years earlier in Warsaw, in a cosmopolitan atmosphere of intellectual and cultural refinement, had every advantage that Hélène was denied. Marie's father, a professor of physics, stimulated her interest in science and encouraged her to study at the Sorbonne in Paris. Hélène Balagny's situation was exactly the opposite. She lived in a hamlet on the edge of the provincial town of Creil, had no father, and no one around her who even knew what it meant to study. Moreover, from an early age, she was expected to take care of herself financially. Even if her family had been prosperous farmers rather than poor peasants, it is still unlikely that she would have had the benefit of a "proper education." Primary school was the limit for provincial girls in France in her era, and the only possible place for further education was at home.

Nevertheless, Hélène did distinguish herself during the limited time she had as a student. She was so far ahead of the

*Jean Théodore Didier van Heijenoort, before leaving the Netherlands, ca. 1905.*

rest of her class that she was asked to take the examination for the *certificat d'études* two years early. She passed the exam with the highest marks and great acclaim, but her triumph was short-lived. Ironically, her brilliance had worked against her and instead of a beginning, her early success meant the end of her student career. The *certificat d'études*, with or without the honors she received, signified (indeed certified) that a girl's education was completed, and that she was ready for work or marriage.

So the day after her outstanding scholastic performance, the petite Hélène Balagny, aged eleven, went to work as a domestic. *C'était normal*, as the French would say, for girls of her economic class and social station. The only unusual thing about it was that Hélène was younger, smaller, and smarter than the average. Otherwise, she was simply continuing a family pattern and doing what her mother had done.

Adèle Louis, van Heijenoort's maternal grandmother, was born in 1848, also in the Oise Valley. She married Monsieur Balagny and had three daughters; the youngest, Hélène, was born when she was thirty-nine. Almost immediately after Hélène's birth, for unknown reasons, Adèle divorced Balagny, took back her maiden name Louis, and went to work as a maid to support herself and her children. This was a remarkable thing for a woman in a provincial Catholic town to do at that time. The law permitting divorce had just been passed (in 1885), but only a rare woman stepped forth to take advantage of it. Adèle, however, was that kind of woman: an extraordinarily forceful person, remembered by her family for her independence, integrity and high standards, and by her grandson Jean for the way she made the town butcher tremble with her demands for the best and freshest cuts of meat. In ringing tones she would say, "Do not give me *frigo!*" (meat imported from Australia that was stored at freezing temperatures for long periods).

If there was no tradition of education in Hélène's family, there was, in its place, that streak of independence and courage. She finished school early, but did not marry until late. Furthermore, she did not marry a Frenchman. At twenty-three, while she was still working as a live-in servant, she married Jean

Théodore Didier van Heijenoort, the Dutchman. Hélène's marriage to a stranger was, in its own way, as bold and provocative a step as her mother's divorce had been, a sharp defiance of convention, and there can be little doubt that tongues in town were set to wagging about those who court disaster.

Gossip aside, there were, in fact, some very good, practical reasons why a Frenchwoman was ill-advised to marry a Dutchman, or any foreigner, unless she planned to live with him in his native land. In France, as in most of Europe, the law, based upon the Napoleonic Code, stipulated that in marriage a woman put on her husband's nationality, rather like a garment. Accordingly, when Hélène Balagny married Jean van Heijenoort of Delft, she lost her French citizenship and was transformed into a Dutchwoman. The fact that the ceremony took place in Creil, in her own ancestral valley of the Oise, and that she and her husband continued to live there was irrelevant. From that moment, she, who was rooted in that soil and had never traveled more than 100 kilometers from her home ground, was required to carry papers designating her as an alien. Had she been Henri instead of Hélène, the situation would have been reversed: she would have had the power to bestow rather than lose citizenship. But Hélène she was, and tragedy was to follow within a few years.

Jan Théodore van Heijenoort, the father Jean would remember only as a dying man, was born in Delft, about 1885. He, too, was from a poor family, one that had suffered great personal loss. His father (van Heijenoort's grandfather), Jan Dirk van Heijenoort, was one of two young brothers who had survived the cholera epidemic that raged through the Netherlands in 1848. All the other members of the van Heijenoort family had died. Jan Dirk was reared and educated in a home for orphans where he was taught carpentry and woodworking. When he came of age, he settled in Delft, became a carpenter, married, and had several children. One of them was Jan Théodore who, when he grew to manhood, would leave his own country and seek opportunity elsewhere.

At the beginning of the twentieth century, a period of general mobility, it was not at all uncommon for young men from families of limited means to take to the road and go abroad in search of work. Holland was poor and overpopulated—in spite of the toll taken by epidemics that had decimated entire families. Jobs were scarce, pay was low, and attractive employment opportunities for a man with ambition and talent but without influential family or friends were almost nonexistent. Such were the conditions that induced Jan Théodore—soon to become *Jean*—to leave his native land.

Between 1906 and 1908 he went to Brussels and stayed long enough to learn some French, but finding no job there that satisfied him, he pushed on to northern France. Perhaps he had intended to go on to Paris, but when he arrived in Creil, a small industrial town, he found exactly what he had been seeking: first, a good job, and later, a good woman. The job was with Fichet Coffre-Fort, a well-known manufacturer of metal safes, and the woman was the exceptionally intelligent and high-spirited Hélène Balagny.

Somewhere, perhaps as an apprentice in a cabinet shop or in one of the many ceramic studios of Delft, Jean Théodore had become expert in the art of fine design. At Fichet Coffre-Fort, he was hired as a painter, to execute the elegant motifs and figures that embellished the best safes of the period. His talent was recognized, his work was appreciated, and he was well paid for his efforts. He had found his niche.

When her future husband arrived in Creil, Hélène's situation was also good. She was employed by the Burtons, a lively and literate English couple who owned a large manor house on the outskirts of town. The Burtons treated their help well and particularly appreciated Hélène's rare qualities. In their home, which by local standards was very grand, she had the opportunity to learn a great deal and in later years, she counted that period as the most interesting and carefree time of her youth.

There is no record of the length of the courtship between Hélène and Jean, or whether she had some hesitation about marrying a foreigner. Nor are there reports of what the Burtons or Madame Adèle Louis had to say. But family lore has it that it was a love match, and that Jean was the one true love of Hélène's life. Paulette Corpion (Hélène's daughter by her second marriage) was almost certain that Hélène met Jean Théodore

*The wedding photograph of Charlotte Hélène Balagny and Jean Théodore Didier van Heijenoort. France, 1910.*

*Hélène, Jean Théodore Didier, and Jean Louis Maxime van Heijenoort. 1913.*

at the manor house. As Paulette later reconstructed the story, the wealthy Burtons would have ordered at least one safe from Fichet, and it would have been just the sort of *coffre-fort* that the talented Jean Théodore would have been commissioned to decorate and domesticate. To make the safe compatible with the Burtons' elegant furnishings, he would have had to make several visits to the manor to consult about proportions and consider the design. Of course, Hélène would have been present. In any case, Paulette concluded, "*Maman* never went to dances, so where else could they have met?"

And so, in 1910, Hélène Balagny and the attractive Dutchman were married. The happy young couple found a comfortable apartment and in July 1912 were blessed by the birth of a baby boy, Jean van Heijenoort. Jean Théodore was the proverbial "good provider," and under those felicitous circumstances, Hélène gave up her job at the Burtons' and devoted herself to her own family.

The idyll was shattered two years later, in August 1914, with the outbreak of World War I and the German invasion of northern France. Within weeks, *les Prussiens,* as Adèle Louis called them, were marching down the main street of Creil. She had seen them before, in 1870, when she was a girl of sixteen, and she would see them yet again, in 1940, when she would say with fear and loathing, "*Les Prussiens sont arrivés encore une fois!*" (The Germans are here. Again!)

The war and the German onslaught caught the citizens of Creil completely by surprise. Both physically and mentally they were totally unprepared, and in panic, the entire population fled, scattering into the surrounding countryside. Because of Creil's strategic position at the crossroads of the main railroad line to Paris, it was the target of enemy bombing raids during all four years of the war. But after the first few weeks of fighting, the Germans did not try to keep an occupying force in the town, and Creil remained in French hands. Although they were still in the center of a major battle zone, the Creilleois returned to their homes and, in many cases, to their jobs. In time, most of the able-bodied men were mobilized into the army and went off to battle.

There was one small but very significant segment of the population of Creil that did not return after the invasion and evacuation. According to the van Heijenoort family, not a single doctor or nurse came back. Some, of course, were called to military duty and some no doubt sought greater security away from the battlefields. Whatever the cause, the civilians of Creil were left without professional medical care of any kind.

Unlike most men in town, Jean Théodore did not become a soldier in the French Army. He was not legally required to serve because he was a citizen of Holland, a neutral country in the war. But there was a second, more urgent and immediate reason, which would have prevented him from enlisting even if he had wanted to: he had gastric ulcers. Exacerbated by stress, the general terror of the invasion and his particular worries about being an alien with an unpopular name, his sickness became worse. Almost as soon as the van Heijenoorts returned to their home after the evacuation, Jean Théodore's ulcer began to hemorrhage. This was the scene that was seared in the two-year-old Jean's memory: his anguished mother, a white basin, and his father endlessly coughing blood.

The story would be told over and over again by every member of the family: how the doctors and nurses had gone and never returned; how no one in the family or among the neighbors knew what to do; how there seemed to be no way to stop the bleeding or the pain; how there was no one to turn to; and most bitterly, how in the mortifying absence of medical help and knowledge, Jean Théodore Didier van Heijenoort died in pain. He had not gone to battle, but in his family's mind, he was a casualty of war.

Before Jean van Heijenoort was of an age to even begin to understand the meaning of his father's death, the nature and the timing of the event and the series of consequences that followed pervaded his consciousness and determined his child's-eye view of the world around him. His first perception of himself was that he was somehow different from other children. In school, he felt he didn't belong. Building on the memory of his mother's desperation and his own fear and helplessness, he was convinced that he, personally, had been singled out and cursed. He felt he had been "chosen by fate to be unhappy." For van Heijenoort, no milder words would do to convey the gloom of his early years.

"My mother was always sad, always dressed in black. Other children had mothers who wore colored dresses, but not mine," he said later. "From the time I was aware of the world and myself in it, I knew there was something wrong, something that should have been different." Considering the enormous number of casualties in World War I, it hardly seems likely that there were no other widows dressed in black in Creil. Yet that is how the child Jean saw it: he was alien, and he and his mother were uniquely unhappy.

Jean-Paul Sartre, also fatherless, wrote about his father's death in a different vein, as a liberating force in his life, freeing him from his conscience. Because Sartre's father died before he, as a child, could be aware of what it meant, Sartre claimed he felt no guilt:

> Dying is not everything. One must die at the right time. Later, I would have felt guilty: an orphan who knows what is happening makes himself feel he has done wrong ... Me, I was delighted. My unhappy situation commanded respect; it was the basis of my importance. I counted my mourning as one of my virtues.

So sanguine an attitude was impossible for little Jean van Heijenoort. He had seen and suffered all the trauma of his father's death, and, with the mysterious logic of childhood, at least on an unconscious level, felt culpable. His *condition* was much worse than unhappy. He thought of himself as *malheureux* in the word's strongest sense: "cursed." As he perceived it, what had happened to him was not mere accident. He had been singled out, by fate, in order to be damned.

After Jean Théodore's death, there was hardly time for mourning. Amidst the chaos of the war, Hélène had to deal with her personal calamity; she had to find a job to support

*Jean Louis Maxime van Heijenoort, not yet four years old. 1916.*

herself and her small child. Normally this would not have been a problem; she had worked before her marriage, and she was an attractive, intelligent young woman. But Hélène's situation was far from normal. She had become a Dutch wife and her husband's death changed her status in one way only: now she was a Dutch widow. As such, she was required to register as an alien, to report to the police once a week, to carry her alien registration book on her person, and to refrain from leaving the city limits of Creil unless she had a special permit for each departure. The fact that she was not a French citizen disqualified her for almost every kind of job she might find within the city except private domestic work. The hysteria that accompanied the war and the reality that the enemy was indeed in and

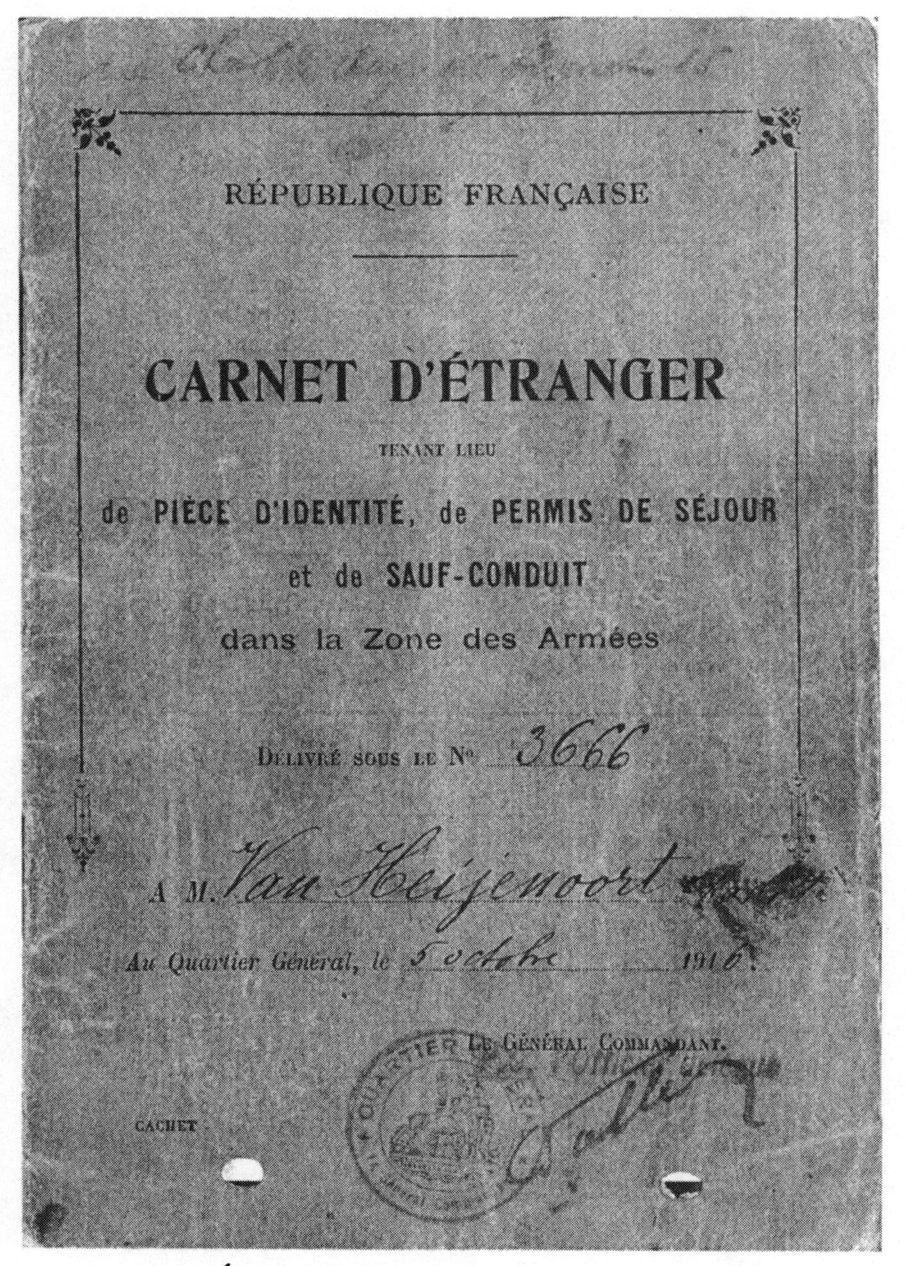

RÉPUBLIQUE FRANÇAISE

CARNET D'ÉTRANGER

TENANT LIEU

de PIÈCE D'IDENTITÉ, de PERMIS DE SÉJOUR

et de SAUF-CONDUIT

dans la Zone des Armées

DÉLIVRÉ SOUS LE N° 3666

A M. Van Heijenoort

Au Quartier Général, le 5 octobre 1916

LE GÉNÉRAL COMMANDANT.

CACHET

*Carnet d'Étranger, Hélène's alien registration papers. 1917.*

around the immediate vicinity made every non-citizen suspect. It hardly seemed to matter that everyone in town knew Hélène. Laws were to be enforced; she was an alien; therefore, she could not be employed in civil service, local or national. She could not get a job as ticket-taker at the train station where her older sister, Angèle, worked as a clerk. She could not be hired to clean the public lavatories because that, too, was civil service. And finally, she was not permitted to cross the bridge over the Oise to look for work elsewhere. Stripped of her birthright, she remained a prisoner in her own town.

The first solution to Hélène's predicament was a place in the Lafeuille household, where her mother worked as housekeeper. The kind Lafeuilles, one of the few prosperous families in Creil, made room for the young widow and her son and allowed them to stay for more than a year, until Hélène finally succeeded in getting a "real" job as a chambermaid and kitchen helper at the Hôtel du Chemin de Fer. Her wages at the hotel were low, but she had the benefit of room and board; and in contrast to her situation at the Lafeuilles', she was able to enjoy a certain limited independence and to have contact with the outside world.

There was, literally, no room at the inn for Jean, and even if there had been, his presence there would have been considered inappropriate. Instead, he was sent to live with his aunt Angèle, also a widow, and her two daughters who were his older cousins. At his aunt's house, on the edge of town, physical space was not a problem—there was plenty of room—but the atmosphere was constricted and gloomy.

By coincidence, the very same week that Jean's father had died, Angèle's husband had committed suicide by hanging himself. There were suggestions that his suicide was linked to the war with Germany and his mortification over his Alsatian origins. All these details, including the nature of his uncle's death, were kept from Jean, but, nevertheless, he interpreted the event as further evidence that he and his family were under a dark cloud.

Except for certain emotional displays, Jean accepted the circumstances of his new situation, but he was miserable. He thought of his aunt as "a good woman and very correct."

*Aunt Angèle Gagneur's house in Nogent-sur-l'Oise, where Jean lived while his mother worked at the Hôtel du Chemin de Fer.*

She was never mean, but she was not at all like his mother. She wasn't warm and loving. She never smiled. He longed for his mother and wanted to be with her, only. A hopeless wish.

The only long visits Jean was able to have with Hélène were on Sunday afternoons, her day off. Bittersweet sessions they were, which almost always ended badly. Each Sunday evening, when it was time for her to leave, he would mourn her departure with a full-fledged, desperate tantrum, flinging himself at his mother to keep her with him longer. In the aftermath of these explosions, he would manifest his misery by silence. Recalling his emotions, van Heijenoort said, "I wouldn't talk, and then, sometimes I would burst into tears and cry for no reason at all, when I was alone and no one was watching. It wasn't for show. I was a very disturbed child."

The realities of Hélène's situation put Jean in a terrible bind. Forces beyond his comprehension seemed to make it necessary for his mother to "abandon" him, week after week, and he was supposed to understand that she had no other choice. She had to work, did she not? And his aunt Angèle was good to him, was she not, so how could he complain? True, he had no father, but his mother and his aunt had no husbands. There was nothing anyone could do about such things. He was expected to be brave and do what he was told – a tall order for a sensitive child.

Van Heijenoort's phrase, "It wasn't for show," carries the implication that there might have been times when his sadness was indeed "for show" and that as a child he was at least partially aware of the difference between dramatizing feelings for public consumption, and private, unarticulated, unconsoled suffering. Much later, more than one person suggested that van Heijenoort as an adult was not averse to playing the role of the injured innocent, particularly in his affairs and friendships with women. Not unlike Sartre in this respect, he did use his unhappy situation to gain both sympathy and respect from the women who surrounded him. And the tactic was sometimes successful, although the attentions he received offered only temporary relief.

In 1919, when Jean was seven, Hélène married Georges Doré, a returned war veteran, the new cook at the Hôtel du Chemin de Fer, and the following year Paulette, Jean's half-sister, was born. It was not until Jean was eleven that he went to live with the new family, and then, as it turned out, it would be just for one year. By that time, the damage of his separation from his mother had been done. "It was too late," he would later say. Her remarriage had compounded the fracture, and the pain to which he had become accustomed sharpened and grew worse. In later years, van Heijenoort made much of his father's death and its effect upon him, but he glossed over the effect of losing his mother. He lost her in two stages: first to her job at the hotel and then – much harder to take – to another man. Jean never said so explicitly, but perhaps one of the worst consequences of his father's death was that his adored mother replaced him with the cook from the Hôtel du Chemin de Fer, a man he could never care for.

*Jean and his cousins during the evacuation of 1918. "Souvenir of the evacuation, Orléans, October, 1918."*

TE POSTALE

Souvenir de l'évacuation
fait à Orléans
le 5 Octobre 1918

Aside from the tantrums of his early days, Jean never allowed himself to acknowledge any anger towards Hélène. "He put his mother on a pedestal. She was perfect," Laure van Heijenoort said of her father's feelings for his mother. "He did the same with all his women at first." But it was only at first. In the case of his wives and girlfriends, when the honeymoon was over and the relationship came down to earth, there was always serious trouble. Jealousy, anger, bitterness, regret clouded all his romantic associations with women. The emotions he had not permitted himself towards Hélène would be released time and time again in the direction of other loves while *maman* remained on the pedestal forever. The only anger he demonstrated was directed towards Georges Doré who he had no desire to accept as a surrogate father. From this point onward, "It was too late," became the leitmotif that van Heijenoort repeated to express his feelings about a hurt which had gone so deep he could no longer imagine it being healed.

The picture of Jean van Heijenoort's tender, formative years is unquestionably bleak, but that is not to say it was unrelievedly so. Although his people were poor, no one went hungry; everyone had clothes to wear and a place to live. But more than that: in the midst of a family stricken by sorrow, he was a brightness, the darling of his "milieu of women," a blonde wonder child, "a brilliant star," according to Paulette, who more than half a century later recalled, with a mixture of pride and lingering resentment, the fuss everyone made over her half-brother. In their eyes young Jean was acclaimed as a genius destined to go forth and shine in territory they had not dreamed of entering. As the only male, he was catered to and encouraged. And although Jean bemoaned being different—not like the other boys—and labeled himself "alien," he would later acknowledge the *pleasure* of being distinctive: "Yes, I always had the feeling of being very special, of being set apart, and sad because of my father; but in a funny way, I have to admit, I liked it."

# Chapter Two

# "Those Schools"

1919–1929

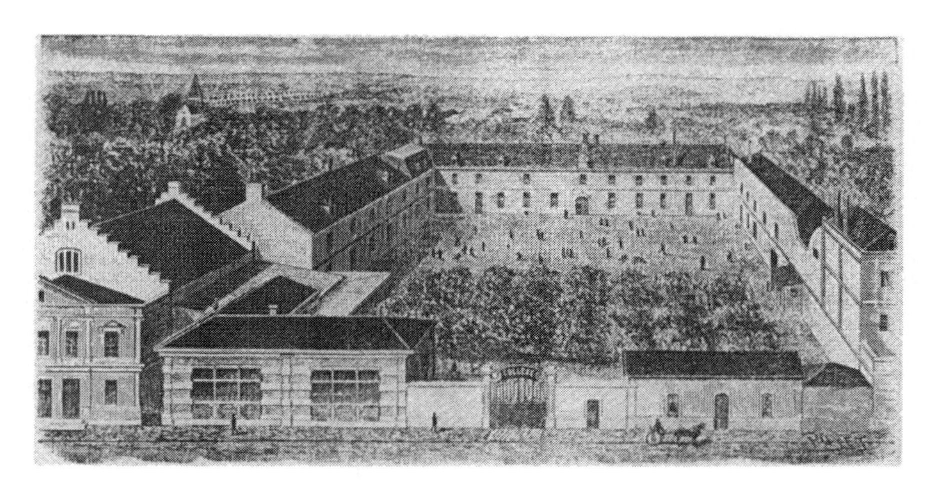

*Collège de Clermont, 1920.*

As a child, I used to hear people talking about something they called *avanlaguerre* [*avant la guerre,* before the war]. At first, I had no idea what that meant. I didn't know if it was a time or a place. It was, Oh, *avanlaguerre* when we had eggs, I used to make a *gateaux St. Honoré,* or *avanlaguerre* I used to have a flower garden. It was like something in a fairytale, because to me those things weren't real. Reality for me was the war, bombs dropping and people dying. Then, when they began to talk about peace, I didn't know what that meant either.

JEAN VAN HEIJENOORT, *1983*

ONE MILLION four hundred thousand Frenchmen were killed in World War I from August 1914 to November 1918, including half the male population between the ages of twenty and thirty-two. These were the years of Jean van Heijenoort's early childhood; this was the war he would remember as his "reality." His recollections of the time are so remarkably detailed and precise, one might wonder if they are truly the memories of a child not yet six years old. But, as in the case of his father's death, van Heijenoort insisted upon the fact that they were indeed his own, and not reconstructions based upon hearsay and later information. There is good reason to believe him: his memory, the display of which never ceased to amaze his friends and colleagues, was prodigious. He was one of those people who remembered everything he ever read and every movie he saw.

In contrast to the deeply personal trauma that Jean suffered as a result of his father's death, his reaction to daily life in the midst of war was detached. From a small boy's point of view, the war even had certain virtues. There was plenty of excitement: Huge tri-color flags were hanging in the streets, and men, newly outfitted in military uniforms, would sometimes give him and the other children rare treats of chocolate and biscuits. Many of these soldier-men were still in their teens, boys really, called up by the year of their class in school. Van Heijenoort later recalled watching the departure of the class of 1917. "They all got drunk and had a great time before they marched away. We didn't know it then, but of course most of them were marching off to be killed."

Because Creil was an important rail junction, the Germans conducted regular bombing raids on the town. On moonlit nights their monoplanes would follow the reflection of the steel rails, flying very low. At 200 meters, according to van Heijenoort's later, and always precise, account they would drop their bombs:

> On those nights, when the bombs were falling close to my aunt's house, my mother–if she was there–or my grandmother or aunt would wrap me in a blanket and we would go down into the cellar. In the morning I would go out to see what had happened, and there would be these big craters with water in the bottom and dead cats floating on top. It gave me a funny feeling but I wasn't frightened.

Recalled from a child's point of view, van Heijenoort's descriptions of war-time incidents were scientific: sometimes ironic and sometimes even tinged with black humor, but devoid of the sorrow and pain that crept into his more directly personal recollections:

> One day I watched a man who was on the road working a steam roller. I saw the German planes coming and so did he, and I saw him get out of the machine and hide underneath it. But they must have thought the steam roller was a big gun, a cannon, because they bombed it and the man was killed. But I don't remember being genuinely scared. It was more like the sensation of watching a movie.

During the second battle of the Marne, in July 1918, when the German armies in a final ferocious offensive came within thirty-five miles of Paris before they were pushed back, the situation in Creil became so much worse that most of the children were sent south to be sheltered wherever homes could be found. Jean and his cousins were taken in by a farm family near Orléans. They stayed until almost the end of the war, saw the grape harvest in October, and got a whiff of farm life which pleased Jean very much. As it turned out, for him, it was an exciting adventure rather than a terrifying experience.

When he went back home, his mother's embrace seemed to mark a turning point. After four years of life in the backyard of the battlefield, peace was just around the corner. But, in fact, for Jean, the end of the war, and the jubilation that followed, did not bring an end to his troubles. What he was subjected to after the war by his local "friends" hurt him as profoundly as the enemy's bombs.

He was six now, and, still living at his aunt's house, he had started primary school, trekking the considerable distance four times a day, French style, in order to have a proper lunch at home. He already perceived himself as an outsider, and he was sure that his classmates also knew that he didn't belong. His name alone was enough to make him feel recognizably apart, but after the German defeat, his situation became worse instead of better. When the Kaiser was given refuge in Holland, the anti-Dutch sentiment in France–already prevalent during the war because of Holland's neutrality—increased. Jean's schoolmates were more than happy, in the spirit of vigorous patriotism, to act out the French attitude. They taunted him, called him *sale Boche* (dirty German) and frequently beat him up for belonging to the nation that gave comfort to the enemy or *was*, in their minds, the enemy. The injustice of those beatings hurt Jean's already logical mind as much as the blows hurt his body. On the one hand, he knew he was not German, and he also knew that at six he was not responsible for the war or Holland's neutrality. On the other hand, how could he be sure about what his father would have thought or said? Here was fuel to be added to his conviction that he was surely among the *malheureux*.

Jean's mother's new marriage compounded his misery. For Hélène, the very obvious virtue of Doré's being French meant that under the same Napoleonic code that had snatched it away, her lost citizenship was restored. If marriage to this solid, *petit-bourgeois* was not like the love match of her first union, there were, nevertheless, many ways in which her life was improved. But for Jean, it was another searing loss. All the reasonable explanations for why his mother would marry again were of no help to him. Nothing about Georges Doré tempted the boy to think of him as replacement for his lost father. Jean was now past the age of a weekly tantrum, and he loved his mother too much to risk overt hostility. The only person he could safely blame and dislike was his step-father, who had taken his mother from him, given her a new name, a new baby–his half-sister Paulette–and, in a few years, would take her to a different city.

Even after Hélène's remarriage, Jean remained with aunt Angèle and his cousins. At first, this was due to the postwar housing shortage. Since many houses and apartments had been destroyed during the bombardment of Creil, the Doré family had been able to find only makeshift living space. Later, Jean was enrolled as an *interne* in his school at Clermont de l'Oise which meant he was obliged to live there. As the son who sometimes came home on weekends and holidays, he belonged to the new family unit only part-time and peripherally. Now, even in his own family, he was the odd man out.

Defining himself as the alienated outsider at so tender an age made it necessary for Jean to create some kind of defense. To this end, consciously or otherwise, he built a protective shell of silence and detachment, a place to retreat. He wore his carapace, but he was not content. He wanted desperately to belong, to somebody, or some *thing* with powerful meaning or structure. Not then, or ever, did he cease to look for community, purpose and engagement.

To van Heijenoort's later bafflement, he first found what he was looking for in the church, during a brief period of religious intensity that occurred while he was living with his aunt Angèle. She was a much more religious and observant person than his mother, or the average Creilleois. According to van Heijenoort, "Even in that provincial Catholic town her faith was thought to be *un peu exaggeré* [a bit too much]." Indeed, this was one of the things that made Jean uncomfortable about his aunt.

As a matter of course, like all the other boys in the community, Jean went to catechism class and became a choir boy. Because of his seriousness and his aptitude for learning anything and everything, he was chosen to help the priest with the ritual of the mass. He did all the things he was supposed to do and more. When he came to assist at the seven o'clock mass, he and the other boys would also have to ring the bells, which required much physical effort. In the winter, the cold rope would rip their hands, but then they would be lifted off

*The Doré family: Jean, Paulette, Georges, and Hélène. 1922*

the ground by the weight of the bells, and that, van Heijenoort found thrilling.

On occasion, all the choir boys would be sent to the country to bring back the coffin of a small child who had died. This kind of assignment was deeply upsetting to Jean, not because he focused on the death of the child, often younger than he was, but because he was reminded of his father, whose death permeated so much of his consciousness. Nevertheless, he never refused or tried to find an excuse not to go.

For a few months van Heijenoort was sustained by what he called "a kind of effusion, a childish mysticism." He remembered taking some chalk and writing *DIEU* in large letters on walls everywhere, which seemed a very extraordinary thing to him afterwards. This intense devotion soon gave way to a feeling of indifference and was then followed by a period when Jean was bent on making mischief with the other boys. He later described his favorite caper, performed during the ritual part of the mass, when his instructions were to pour a little wine and a drop of water into the chalice the priest was holding:

> I would always try to put in more water and less wine. The priest couldn't say anything, of course, but with his little finger, using all his force, he would exert pressure against the water bottle so that I couldn't add any more.

At age eleven, shortly before he began his long years of boarding school at Clermont, van Heijenoort came to the conclusion that he no longer believed in God and that Science explained everything; or, if it didn't yet, it would. With no one guiding him or arguing with him one way or the other, he "simply ceased believing." The change came about abruptly, but it was deep and irrevocable.

Dropping everything religious pertaining to God or a supreme being, van Heijenoort reapplied his zealousness to his studies at the *collège*, "absorbing knowledge" like a sponge and looking for new answers. For those years he belonged to the cult of learning and scholarship; to his studies he transferred all the passionate attention he had previously given to God. Later, in Paris, at the Lycée Saint-Louis, he would continue in this vein until he was overtaken and smitten by yet another creed to which he made an all-consuming commitment, this one allowing room for nothing else. In this total devotion to Marxism and Trotsky he was to find, at last, a sensation of belonging and comradeship he had never experienced before.

From almost the moment he entered primary school in 1918, van Heijenoort knew he had found a world which would bring him peace and give him pleasure. It was a stable, orderly realm in which he was soon very sure of himself and his abilities. If he suffered elsewhere, in the classroom he thrived and became a phenomenally successful student. For the rest of his life, education and scholarship of all kinds would be a haven for van Heijenoort. Surrounded by books, immersed in thought, he could create an island of calm or excitement; it was the one environment he could control, at least to some degree.

In terms of intellectual life, Creil was a backwater. Nevertheless everyone in the district, rich or poor, had the benefit of an elementary education. Since in France the educational system was and is a national one, the children of Creil, Paris, and Lyon were served the same curricular fare until they were eleven years old. Then they were to choose, or more likely have chosen for them, a *métier*. Most boys and some girls spent the next two years in a trade school after which it was time to go to work. Each year however, there were a few bright boys who were urged to continue their education in the classical mode and prepare for the baccalaureate examinations. Creil did not have a secondary school, but it was possible to go to the *collège* in Clermont de l'Oise, the sub-prefecture of the district, fifteen kilometers away.

To the credit of the French educational system and the teachers of that period, brilliant boys like Jean van Heijenoort were encouraged and stimulated to think beyond the confines of their provincial town. Even so, being recognized as an outstanding student was by no means an automatic guarantee that a boy would continue with his academic work. A boy's parents had to agree, and not all parents would do so. A poor family often needed or wanted their children to go to work and contribute to the income of the family as soon as possible.

Of course Hélène, the scholar *manquée*, gave her consent. Year after year, her son had been ranked first in his class in the local primary school, where every child was ranked in every subject. And each year, Hélène and all the other women in the family had been there to watch him receive the *prix d'excellence* in several subjects, and to accept the congratulations of their friends and his teachers. She was gratified and proud that her child was gifted in this way and that the opportunity for a serious education, which had been unavailable to her, was open to him. At last, here was something for her troubled boy that seemed to be all good.

In the spring of 1923, accompanied by the teacher who had been his mentor, Jean van Heijenoort traveled by train to Beauvais, the prefecture of the district, to take the competitive scholarship examinations for the whole region of the Oise. The trip made an indelible impression on him. He spent an entire day answering questions in mathematics and other subjects and had to write a long *dictée.* The result was good: he had placed third and was granted the scholarship known as *bourse complète* which assured his entry into secondary school.

In van Heijenoort's year, only four boys from Creil took the academic route to the Collège de Clermont, where they were henceforth to be known as *les quatre Creilleois.* Besides Jean, there was the schoolteacher's son, the pharmacist's son, and Roland, Jean's closest friend, whose father was a railroad engineer. The others were from families who could afford their children's education, while Jean's mother was willing and able to let him go only because of the prized *bourse* which would pay for his room, board, and incidentals at the college for the next seven years. The *bourse* was indeed *complète*: incidentals included books, pencils, paper, laundry fees, and shoe polish.

Clermont de l'Oise, a smaller town than Creil, was important because it was the sub-prefecture of the district, and the academic secondary school on its hill added to its lustre. Only fifteen kilometers away, less than a half-hour by train, to Jean it was a different, unfamiliar world and he preferred his own. "A very reactionary little place," was his disdainful assessment, "much more bourgeois than Creil." He would have preferred to live with his mother or even to continue to stay with his aunt Angèle, but he was not given a choice or even a voice in the decision.

In 1923, a few months after Jean began what he called ever after his "incarceration in *those* schools," his mother, stepfather, and baby Paulette moved to the outskirts of Saint-Quentin, a city 100 kilometers away, to open a café-restaurant of their own. Hélène decided that her son would remain in Clermont where he was already settled as scholarship student and *pensionnaire.* It was a decision Jean fought, but to no avail. Once again circumstances beyond his control took his mother away, and this time the distance seemed enormous. He felt more cut off than ever before. He wanted to move with his mother, even if it meant living with his step-father, and did not understand why he could not. He would not accept the reason that his scholarship was not transferable to the *collège* in Saint-Quentin. He was sure that it could have been arranged. Arguing that he would get a better education in Saint-Quentin because it was a larger, more important city, he tried to convince his mother to take action. But Hélène made no attempts to look into the matter, telling Jean that the amount of paperwork and bureaucracy involved would in the end make it impossible.

With decades of hindsight, other members of the family offered different explanations for the decision to leave Jean behind. Van Heijenoort's half-sister Paulette said,

> There was no question of his doing his studies where we were. We were in a village across the street from a silk factory, two kilometers out of town. After the war you had to be like pioneers to go live in a place like that.

Jean van Heijenoort's son, who came to know the family history well, made the further point that during that period his grandmother and step-grandfather "worked like slaves from five in the morning until ten or eleven in the evening, every day, with no vacations. They had no time to do anything except work." In the younger van Heijenoort's opinion, this was the major factor in the decision.

Hélène herself put another perspective on the matter. Years later she confided, not to her son but to her daughter-in-law Loretta Guyer (Jean's second wife), that she had never ceased to regret leaving Jean behind. She had done so because besides the hardship of their living and working conditions and the impossible bureaucracy involved in a transfer, she thought it was a good idea to keep her new husband and her son apart.

Relations between Jean and his step-father were always strained and problematic and since her son seemed to be doing well at Clermont, why not let him stay where he was? But afterwards, she felt it had been a serious mistake. Van Heijenoort's reaction to this was to deny his dislike of his step-father and to contest his mother's explanation. At the same time, while defining how he felt about Doré, he said, "I just never had any warm feelings towards him. He wasn't my father; he never was like any father; and anyhow it was too late."

As with all the other separations in his life, van Heijenoort had to accept the situation, to make the best of what he thought was a bad bargain; so he did, with resignation. Except for those early childhood tantrums, he did not rant and rave. But appearances can be misleading; van Heijenoort did have a means of retaliation. A knife-like, chilling silence became his weapon, a powerful tool which he used skillfully to show his deep resentment and to make those close to him uncomfortable. His was a double-edged stoniness: by being generally unresponsive, by appearing disinterested and by refusing to talk, he protected himself, but he also made manifest his pain and his contempt. Under the guise of natural shyness, he found a method of striking back that suited him perfectly.

Once it was clear to Jean that he was not going to be transferred to the school at Saint-Quentin, he plunged into his schoolwork in Clermont. There was no ambiguity and almost no complaint in his assessment of the quality of the academic education; for a bright student, it was excellent. His only qualification was that the curriculum was too fixed. There were courses, lectures, textbooks and homework but no free choice and very little time for outside reading. At almost the same time, students all over France were reading La Fontaine, conjugating *amo, amas, amat,* and taking the square roots of hypotenuses. Although the program was indeed inflexible, it had undeniable virtues, since it meant that country boys were not shortchanged. What differences there were in schooling had to do with the instructors and not the prescribed content or the texts. And

*Jean (second row, sixth from left) at la colonie de vacances (summer camp). 1923*

Jean thought his instructors were very dedicated, very good, and, sometimes, even excellent.

On the other hand, the physical and emotional atmosphere of the school was something horrendous for the sensitive van Heijenoort. In earlier times, the Collège de Clermont had been a convent, built in an austere spirit and not designed to cosset its inmates. Van Heijenoort had a long list of never-to-be-forgotten complaints and grudges which stemmed from injustices, some petty, some cruel, that he was powerless to rectify or avenge:

> The discipline at that school was absurd—and the meals were terrible. A bell would wake us at 5:30 in the morning. There was no heat in the dormitory, so in winter the water in the toilets would be frozen. We had to wash in freezing water, too. Then we went to a room called *la cordonnerie* where we had to shine our shoes before we could go down to the refectory. Breakfast was a watery soup and a few slices of hard bread and since everything was arranged like clockwork, the *garçon* had to pour

> the soup before we came in, so naturally it was cold. Imagine, in winter starting a long hard day on that cold soup. But, worse yet, students whose parents had paid a supplement of twenty francs per month were allowed to have *café au lait* or hot chocolate in the morning. Of course my mother could not afford that, so while I was having the lousy, cold soup, the guy next to me was having a steamy bowl of delicious hot chocolate which I could smell. That was the cruelest thing. Much later I discovered why the food was so bad. The principal of the *collège* was given a flat fee for every *pensionnaire*. By skimping on meals he was able to pocket a little extra money.

The "absurd" (one of van Heijenoort's favorite pejoratives) discipline in the school was enforced by a group known as *les surveillants*, young men, usually university students, who took the job for the small income it brought. Some were sympathetic but others were petty or outright sadistic types:

> If we got the least bit noisy while we were eating, they would shout, "Silence!", and then we would have to remain silent for the whole rest of the meal. Or in study hall, if you dared to speak to your neighbor, they would shout, *"Vous me copierez dix fois La Fable des Rennes"*. [You will write out the Fable of the Reindeer ten times.] It was sheer sadism. *La Fable des Rennes* has 400 verses! On the other hand, there were no beatings. Never. Later, when I read about the canings in the British schools, I was absolutely shocked. That form of punishment was unheard of in French schools.

In spite of all his complaints, van Heijenoort did more than simply adjust to life in the institution. He would never praise the whole, but he did reluctantly acknowledge an active and genuine liking for many aspects of the daily routine at Clermont. In a sense, it was an acceptance of himself, for so much of the *real him*—his small habits and reactions, his set of mind and way of thinking—was formed there. The rigor of the work, the strict order of the day, these were not at all antipathetic to his nature, and the fact that he always knew what he was supposed to be doing at any given moment and could count on what was scheduled to happen next gave him a feeling of security that was absent in the other areas of his life.

He also developed an acute sensitivity to the social interactions taking place around him and developed his talent for watching without being noticed. "It is very subtle what goes on in an institution," he later said. "There are these small groups where you become very close. Your life and your relationships can be extremely rich." For Jean, this intense camaraderie with a few intimate friends took the place of family.

From the first to the last of his seven years at the *collège,* van Heijenoort was acclaimed for his brilliance as a student. For him, the personal rewards of being acknowledged in this way, privately and publicly, went far beyond their academic importance. There was the delicious moment when a teacher or a surveillant would call him aside to whisper the secret, *"Vous savez, vous avez le prix d'excellence"* (You know, you have won the prize for excellence). Later, there would be the theatrical award ceremony where the names of the winners were announced, supposedly to everyone's surprise. His name would be called; he would go up to the stage and shake someone's hand and receive the award. Walking back to his seat, his arms loaded with books, he would look over at his mother, who on these occasions was always in the audience, and smile at her. By the end of his college days, van Heijenoort had a good start on his personal library, for he received as many as twelve books a year on these occasions.

In Clermont, his energies and capabilities seemed to be boundless. By one of life's amazing coincidences, when Dr. Françoise Ville, later a faculty member at the University of Paris, was a little girl, she used to hear her father speak about a most outstanding person, a boy he once knew at the Collège de Clermont, who had not only won the *prix d'excellence* in every subject, but also wrote plays in Alexandrine form and then produced and acted in them for audiences at the school. Dr. Ville's father continued his story with the information that this unusual person had later become Trotsky's bodyguard and secretary, and after that, believe it or not, he had become a logician. Years later, when Ville herself became a logic student, she inevitably came across the name "van Heijenoort." When she heard the ever-circulating rumors of Professor van Heijenoort's past, she reasoned, correctly, that there could not be two such individuals.

Another dramatic event of broader public significance took place during van Heijenoort's last years in Clermont. In response to the agitation and growing pressure that had been brought to bear upon the school system, in the late 1920s, girls bent upon going to the university were finally admitted to the provincial *collèges* so that they, like boys, could prepare for university entrance examinations. It was no longer a cut-and-dried situation that a young woman's education was complete at thirteen. The time had arrived when young women with the ability, drive, and desire for further education had at least a small chance of getting it. The Hélène Balagnys of France now had the *possibility*, if not the probability, of becoming physicists—provided they had their parents' agreement.

There were just a few girls who came only for their classes and then disappeared, but the effect upon the boys was sensational, van Heijenoort recalled:

> We went about whispering to each other, saying *"Vous savez, les jeunes filles au collège!"*. We could hardly believe it. Then, almost immediately, all sorts of intrigues began and by the time I was in my senior year, one of those *jeunes filles* became my girl friend.

Jean was timid — "I am terribly shy," he would say — but his self-proclaimed reticence never seemed to hamper his success with women. A few girls and hundreds of boys, yet *he* had a girl friend. How? His almost blushing response was, "Well ... she took a certain initiative." The pattern was to continue for the rest of his life. There was never a shortage of women happy to take "a certain initiative" towards the tall, handsome, and locally famous van Heijenoort. It was also true that the smallest, subtlest sign from someone he found attractive caused him to overcome his terrible shyness in suprisingly short order. His experience of growing up with his "milieu of women" stood him in good stead. He was comfortable and at ease with female friends of any age in a way that he rarely succeeded in being with men, and it wasn't merely a question of romantic interest. He simply liked women and their company.

*Les jeunes filles au collège* was one kind of revolution. There was, at the same time, another very personal revolution taking place in the minds of a few students at the Collège de Clermont, which affected the way Jean was to define himself and the world around him. When he was fifteen, he was recruited into a group of six or seven boys who met regularly to discuss "urgent issues." Their leader and organizer was one of the *surveillants,* "not a mean one–a nice one, a university student who knew the score." By "knowing the score" van Heijenoort meant the student-leader had the correct slant on the political scene and, in particular, on Marxism. The group was extra-curricular and secret; the *de rigueur* basic reading materials were smuggled copies of *L'Humanité,* the official Communist Party newspaper.

Danger, intimacy, secrecy, even conspiracy—the boys thrived on it. Jean, the outsider, now saw himself as an insider with privileged information. Now he, too, knew the score. He thought of himself as a "communist" but the *surveillant*-leader was a Trotskyite, and not a Communist Party member. The boys were sure that if the principal found out about their group, they would all be thrown out of school. Their leader would certainly lose his job, and even the boy who smuggled in their newspapers, but was not a member of their club, was at risk. Nevertheless, the group continued.

These clandestine meetings marked the formal beginning of van Heijenoort's rebellion. From his reading of Rousseau and Voltaire he already had developed ideas about freedom, justice, equality, and brotherly love, but those were books he had read in his French literature courses where the aesthetics of the work was given at least as much importance as the philosophy. By contrast, the secret Marxist society was definitely *not* literary, and the young comrades thought they were dealing with politics, not style. Although he was still a teenager, and not yet the person he would later describe in his memoir of his years with Leon Trotsky as someone "in total revolt against society," he was well on his way.

Fed by his early deep experience with misery and injustice, Jean developed a personal interest and a passion for justice

*Jean at twelve, in his Collège de Clermont uniform, with his half-sister Paulette. 1924.*

and fairness that went far beyond intellectual exercise. His conception of war was also formed by personal experience that extended far beyond the four years of war. At eight and nine, Jean and his companions had gone exploring in the neighboring countryside, looking for shells, bullets, guns, cartridge belts, shrapnel–the detritus of battle. Wandering and playing in fields where the soil had been soaked in blood, where the dirtiest fighting of the war had taken place, they found bones and parts of bodies. It was a macabre game, but the reality of death was familiar, and not as consciously distressing to him as the abstract representation of death that he would later see.

A few years after Jean's mother and stepfather moved north to Saint-Quentin, his mother presented him with a rare treasure for a poor boy: a *Diamant Français* bicycle with elegant wooden-rimmed wheels. He kept it in perfect condition and soon gained the confidence to ride his bicycle instead of the train on vacation visits to his family. It was his first taste of independence, and, at a tempo he set for himself, he rode all the routes of northern France, in almost the exact paths of the old battle lines. That experience gave him yet another perspective on the effects of war.

When Jean saw the miles and miles of little white crosses in the endless cemeteries of northern France, he began to comprehend the extent of the devastation. The vision haunted him; he was unable to put it out of his mind:

> I saw the military graveyards. You have no idea what that can be like ... small white crosses as far as the eye can see in any direction. You cannot believe it. It is too impossible, too absurd. Whether the dead were French or British or German didn't make any difference to me. It was the *numbers*, the endless rows, that made such a deep impression. I know it contributed to my political orientation later on. I felt I had to reject–totally reject–the system that caused the war. That was not my kind of world, I decided. I wanted something else.

Small wonder then that van Heijenoort was ripe for recruitment into the secret society of budding revolutionary idealists who were also ready for "something else." Always the scholar, Jean threw himself into the recommended texts with the same intensity he gave to more traditional scholarly stuff. He read Marx and Lenin and, more than incidentally, Trotsky's early autobiography, *My Life*. Although he was reading for content, the vividly powerful literary style of Trotsky's book affected van Heijenoort deeply; he felt an immediate personal identification with Trotsky, the man, and a strong attraction to his ideas on international revolution. On both an emotional and an intellectual level, he was ready to be convinced that Marxism was the solution to the world's ills.

He was fifteen when he came to the momentous decision which would, in a few more years, drastically change the di-

*Jean visiting the family at the café-restaurant, A La Maison Doré in Saint-Quentin. 1927.*

rection of his life, and, not incidentally, his personality. He discarded his earlier perceptions of himself as alien and destined to be unhappy. Instead, as he put it, he was "overtaken by optimism." He thought not only that it was possible to change the world, but also that *he* would play a part in the event.

It was a grand if not a grandiose notion for a timid person, but, in fact, it was not so out of line at the time. Such things were in the air in postwar France. The depressed economic situation and the climate of political and social unrest had created an atmosphere favorable to revolutionary ideas, and a large portion of the thinking population was generally sympathetic to proposals for radical change. As for van Heijenoort personally, fate had loosened its stranglehold on him. At Clermont, he was no longer the child singled out for misery. Quite the contrary, he was the one chosen to be among the best. He had been recognized as a young man of outstanding capabilities, and he had acquired confidence in his intellectual power. He was ready to take on society's evils and put his talents to use for the greater good of mankind. All he needed was a leader to show him the way.

*Jean (front row, center) holding the ball, Collège de Clermont. 1930.*

*Jean (second row, far right), Collège de Clermont. One of the rare jeunes filles in front row. 1930.*

Van Heijenoort's new passion for political ideas and his romance with the *jeune fille* occupied him some of the time, but his desire and his ability to concentrate on academic subjects remained unabated. In his last year at the Collège de Clermont, in a miraculous frenzy of activity, surpassing anything he had done before, Jean studied for the baccalaureate examinations in mathematics and philosophy. These were the exams taken by every secondary student in France, and it was extremely unusual to attempt to do two subjects. After the mathematics exams, the principal of the *collège* called him into his office and behind closed doors ceremoniously announced, "I have gotten you a *bourse complète* at the Lycée Saint-Louis." He felt, perhaps with some justification, that he was a partner in van Heijenoort's achievement, and that his student's acceptance to and support from the most prestigious school in Paris reflected as much glory upon him and his institution, for having nurtured the star, as it did upon the young scholar himself. However, van Heijenoort was the only student from Clermont to be accepted at the Lycée Saint-Louis that year, or any other year, as far as he knew, which perhaps gives a better indication of how the credit should be apportioned.

Usually van Heijenoort treated his academic success in a rather off-hand, understated manner, partly out of a seemingly inbred modesty (he abhorred bragging in others and was never guilty of it himself), and partly because he simply took it as a matter of course that he would excel in that way. But his acceptance and the accompanying *bourse* at the Lycée Saint-Louis were of a different order of magnitude, and he was very excited at the prospect before him. It meant an entrée into the big league of academic studies in France, the intellectual equivalent of Cambridge and Oxford in England or any of the Ivy League schools in the United States.

Of equal if not greater importance, going to Paris meant that the newly politicized scholar would be living in the center, instead of on the fringe, of all the important political activity in France. He already knew that the formal academic opportunities would be outstanding; as for the rest, he hoped and expected to find out where and how he would fit in.

## Chapter Three

# A Sharp Left Turn

### Paris: 1930–1932

*Architect's rendering of the Lycée Saint-Louis in Paris*

I thought I was joining the staff of the World Revolution on its wings.

JEAN VAN HEIJENOORT

THE LYCÉE SAINT-LOUIS, on Boulevard Saint-Michel, was and still is one of the most prestigious of the French undergraduate schools. Founded in 1280 as the Collège d'Harcourt by the Canon Raoul d'Harcourt, who was counselor to Philippe le Bel, it has stood solidly facing the Place de la Sorbonne for seven hundred years. Racine, Diderot, Talleyrand, and thousands of others among the intellectually elite have marched through its heavy bronze doors to be enriched and intimidated by the rigors of the education offered within.

The dark old stones have been blasted clean since Jean van Heijenoort's days there and the building has been rebuilt and remodeled. Physically the school has lost its forbidding aspect. Yet it takes only a small stretch of the imagination to comprehend the mixed feelings of the eighteen-year-old boy fresh from the provinces.

Academically, Jean was completely confident. How could he not have been, given his rigorous preparation and record of excellence at the *collège* in Clermont? Besides his major studies in mathematics and philosophy, he had taken chemistry and physics, seven years of Latin, four of Greek, and four of German, as well as French history, literature, and language. Independently, he had read widely, and he had even taught himself Russian, because he had a vague notion that, one day soon he might go to the Soviet Union for a visit.

But socially, young van Heijenoort was, in most respects, a naive country boy. His head, fit to burst with so much knowledge, was attached to a body with legs only recently accustomed to wearing long trousers. It was his mother who packed the big trunk with all the items on the list given to new *pensionnaires*, instructing them to bring: "two towels, one soap, six shirts..." and it was she who accompanied him to Paris by train and then to the *lycée* by taxi to help him unpack and settle in.

Jean had come to Paris ready to take the next big step toward an academic career. There was no student at the *lycée* more prepared than he was to devote himself to scholarship. But he had also come ready to expand his experience, to learn everything he possibly could of the "ways of the world," about which he professed total ignorance. And finally, he expected to find a political movement to fit his Marxist-socialist ideals—something with larger goals than his clandestine group in Clermont. When, to his dismay, he learned the rules and regulations that he as a student was obliged to follow, he was outraged. He thought it "absolutely absurd, ridiculous," and for the rest of his life he referred to Saint-Louis as the "prison where I was locked up for two years."

The inmates of the "prison" were either *internes* (like Jean) or *externes*. The *internes* were, indeed, locked in at night and subjected to frequent bed checks, while the *externes*, who lived at home, came to school in the morning and left at the end of the day. Naturally, the boarders were pained at this injustice. Although everyone shared the trials of discipline during the day, at night, the *externes* were on their own to do as they pleased on the joyful, sinful streets of Paris. By contrast, for Jean and his fellow "prisoners" immured within the dark walls, freedom was a once-a-week affair, granted on Sundays from eight in the morning until eight in the evening, and subject to recall for bad behavior.

Nothing could have been more maddening to the young van Heijenoort. He had just gained his first grand taste of freedom upon leaving the Collège de Clermont. On his bicycle he had wandered at will, covering hundreds and hundreds of kilometers on the backroads of northern France. And now, ready to discover "life" in the big city, living on "Boul' Mich" in the heart of the Latin quarter, surrounded by bookstores and cafés full of attractive people, where was he? Behind the locked doors and grilled windows of the *lycée*.

He was frustrated, he was resentful, but he was realistic. There were lectures he wanted to hear, political meetings he wanted to attend, movies, plays, and concerts, but he held his disappointment in abeyance and did not let it get in the way of his work or his immediate goal. To external appearances he was a willing, even eager inmate, schooled and habituated by past experience to the rigors of the system. In truth, he *was* proud to have been awarded a scholarship at the illustrious Lycée Saint-Louis and was keenly aware of the privilege he enjoyed. And, as at the Collège de Clermont, he was at least comfortable with, and perhaps comforted by, the order and strictness of the daily routine. It provided a kind of security and reliability that suited his already well-developed taste for precision. After dispensing with his litany of complaints about the reform school atmosphere, he always spoke of his education there as "excellent, excellent, no question about that."

Then, and ever after, van Heijenoort had a reverence for rigorous scholarship and a good piece of work of any kind. His highest praise was, "That's very deep." He scorned the superficial: "Oh but that's obvious; we already knew that." This last he would say with his shoulders lifted, palms up and fingers spread open. His own work was never shallow. Large and lengthy projects did not intimidate him; in fact, he embraced the routine and discipline necessary for such undertakings. He counted this capacity as a virtue, but at the same time, he worried about that meticulous, pedantic side of his character and, because he admired flamboyance in others, wondered if people might think him drab.

Cutting across the categories of *internes* and *externes,* there were two other groups at the *lycée*: *les taupins* (the moles), the students in the special mathematics program, and *les cagneux* (the lazy dogs), the students in letters. Van Heijenoort complained about the extravagant amount of work inflicted upon the mathematics students and the fact that the proportion of classes devoted exclusively to mathematics was "totally skewed": three or four hours of instruction every morning. Even on Thursdays, the traditional day off, *les taupins* had to attend what was familiarly called *la planche* (the blackboard). On these occasions, teachers from the other *lycées* in Paris came to interrogate them. At random, a student would be called to the blackboard and the examiner would say, "Monsieur, write this problem and then solve it." On the spot, in front of everyone, knowing full well the disdain and sarcasm that awaited his mistakes, the trembling *taupin* had to try to come up with a solution.

Van Heijenoort rarely made mistakes; he was very much in command of his abilities; and while he never stopped complaining about being overworked, there was that side of him that would not have been satisfied with any less. He was eagerly absorbing knowledge, but rather than being sated, he found himself stimulated and always ready for more of the huge portions of intellectual vittles served up by the academy. The only thing missing was some significant activity beyond school work but that, too, would begin soon enough.

In *Three Who Made a Revolution,* a biographical history of Lenin, Trotsky and Stalin, Bertram Wolfe describes Leon Trotsky as a young student. He could just as well have been writing about Jean van Heijenoort in his first year at the Lycée Saint-Louis:

> The star pupil was never late, did his homework assiduously, was quiet at his desk, listened attentively, learned quickly and precisely, answered—with invariable attention to what was expected—all the questions put to him; wrote well, recited well, easily led his class; was respectful to his teachers in the classroom, bowed deferentially when he met them on the street. Yet even so proper and bright a boy came into conflict with the regime.

Years later, van Heijenoort's response to the above characterization was a smile and a correction: "Well, yes, it could be me, but I always gave the answer I thought was correct, I mean, the truth, even if it was not what was expected. And I didn't bow." Otherwise, he agreed that as students, he and Trotsky were cut from the same cloth, and although the stimuli were different, both were most certainly predisposed to rebellion. Trotsky's career as a full-time revolutionist began in 1897 when he was eighteen. Van Heijenoort waited until school was out and he was twenty.

* * *

In Paris, at the end of the academic year 1930–1931, there were, as always, the ceremonies where distinguished educators awarded prizes to outstanding students. For the prestigious Lycée Saint-Louis no one less than a minister from the department of education would do to make the presentations, and the ceremony itself had the additional cachet of being held at the Sorbonne. The tall, fair Jean van Heijenoort, with the odd Dutch name, no doubt created something of a sensation in the audience that gathered. His first year had been a stunning success. He had been ranked first in his class in almost every subject, and time after time he was called to the podium to receive the *prix d'excellence.* Although in the past, these end of the year accolades had become almost routine for Jean, this, after all, was Paris, not the provinces. Receiving an award from a minister of education in a ceremony at the Sorbonne was a tangible sign of talent of a magnitude he was not yet quite accustomed to.

As for Hélène Balagny, it was the proudest moment of her life. She had come from Saint-Quentin to witness the recognition of her son's achievement. Here, with solemn ceremony in august quarters, she experienced the vicarious fulfillment of her own thwarted aspirations as well as the confirmation that she had done the right thing by sending young Jean off to boarding school. And this was only the beginning. It was clear that her son was a rising star in the academic world, and there seemed to be few limits on how far he would go.

But, against all expectations, the summer of 1931 would be the last time Jean would be named first. There would be no more moments of complicity between Hélène and her scholarly son, no more looks exchanged as he walked off the stage with his arms full of the books awarded him for his accomplishments. In fact, to Hélène's disappointment and chagrin, very soon he would no longer be that scholarly son. He would leave the path he had followed so diligently and make a sharp turn in quite a different direction.

The weekly Sunday furlough granted by the administration of the Lycée Saint-Louis allowed the *taupins* and the *cagneux* twelve hours of freedom. That was never enough time for Jean to seek out all the things he had imagined and hoped Paris had to offer to an inquiring soul, avid for entertainment as well as enlightenment. All week long, in the corridors and study hall, boys would confer in whispers about which movie or play was the one they *had* to see. It was during his escape from the confining quarters of the *lycée* that van Heijenoort developed his lifelong love for film. "I saw all the important movies of the time," he recalled later, "the good German ones like the *Blue Angel* and of course all the Chaplin films." He even managed to see the little tramp in person:

> Chaplin, we all loved him. He came to Paris and I went to the Gare de Lyon to greet him. There were thousands of people, and, because it was mostly a leftist crowd, there were police everywhere. I managed to climb a pole so I could see. At one point, either to protect him from the crowd, or simply to get him on his way, four gendarmes picked Chaplin up, and holding him out flat, as if he were on a stretcher, they carried him out of the station. It was just like a scene from one of his movies.

Van Heijenoort always began his free Sunday in one of the many bookstores on Boulevard Saint-Michel. His favorite was La Librairie Picarde, "because they had all the surrealist literature." Picarde also had a big pot-bellied stove in the middle of the store and wooden floors that had been washed until they were white. Jean found it a homey place, and he would be there at the very moment the store opened. Carefully, he would chose a book, take it over to the stove, and stand there reading for an hour or two. As far as he could remember, he never bought a single book—he simply could not afford it—but no one in the store ever bothered him or complained, even when he peeked between uncut pages. Unlike the Chaplin movies, surrealism was not to everyone's taste. There was no group of aficionados at the *lycée* discussing the sallies and thrusts of that movement. André Breton's and Louis Aragon's latest manifestos were not hot topics in the study hall of *les taupins*, but van Heijenoort found their ideas a breath of fresh air. He also developed an interest in the surrealist painters; Yves Tanguy and Max Ernst were among his favorites, but not Salvador Dali. Later, explaining his attraction, he said,

> French intellectual life at that time was very stifling, but the surrealists had some new ideas: the use of the dream, the free creation. I thought it was all very interesting and appealing. And besides, at that time the surrealists were very political; they had strong reactions against war, against the French politicians. The feeling of social protest in their work, the breaking of old forms, all that merged very well with my political views.

After his mornings in the bookstores, van Heijenoort might go to a play, a movie, or a concert, in that order of preference. He was also inclined to take long walks along the Seine, while he meditated, alternately, upon the state of the world and mathematics problems. In Montmartre, he discovered Charles Dullin's avant-garde theatre, l'Atelier. The regular seats were expensive for a poor student, but in the last balcony, there were wooden benches and the price for those places was only two francs.

Because of the Saint-Louis early curfew, the performances Van saw were always matinées. At a few minutes before eight, all the Sunday adventurers would gather across the street from the *lycée*, at the Bar de la Sorbonne, "*le Barso*," they called it. There they would stand, until precisely one minute before eight, smoking and collectively inhaling their last free breaths while they detailed their day of liberty. Then, with a final gesture of protest, they crossed the Boulevard Saint-Michel in a noisy flock, stopping traffic as they went, and entered the bronze doors for another week in the academic prison.

* * *

The directors of the Lycée Saint-Louis were right in thinking that freedom is a dangerous distraction. During his second year of school, at his favorite theatre, Jean van Heijenoort was to meet two of the most important people in his life: Gabrielle

Brausch and Yvan Craipeau. The former would become his wife, and the latter would lead him to his great hero, Leon Trotsky. Jean met Gaby in the ticket line for the two-franc seats, and a few weeks later they both watched Yvan and his group stage a protest against l'Atelier's current play.

Mademoiselle Brausch was Jean's "type," a lively, slender brunette. On the Sunday they met, she was reading *L'Humanité,* the Communist party newspaper, while she waited in line. Jean stood behind her, reading over her shoulder and smiling. She could not help but notice his tall, blond good looks each time she half-turned towards him. "What are they saying about...?" was all he needed to ask in order to start a conversation.

She was twenty-one to his nineteen, an "older" woman, a real *parisienne,* and on her own. Her father, too, had died when she was an infant, and she had been raised primarily by her grandmother. Now she lived alone in Belleville, the working class district of Paris, and had a job as a bookkeeper. He was in his second year at the *lycée,* locked in his dormitory except on Sundays, and wanting to do something besides mathematics. It took hardly any time at all for them to discover that their two political hearts beat as one, and very soon they became steady friends.

After their rendezvous at l'Atelier, they walked all over Paris, looking like any other entwined young couple as they strolled along the banks of the Seine. Undoubtedly some of their conversation fit the conventional romantic picture, but, because they were serious young people with Marxist-utopian ideals, deeply concerned about the troubled world of the 1930s, much of their talk was about the politics of the Left. As their friendship developed, their discussions focused on the actions one could or should take to further his or her political ideals. Then, as it turned out, within a very short time, van Heijenoort was presented with an opportunity for action which went far beyond anything he had previously thought of doing.

Coincidentally, on another Sunday afternoon at l'Atelier, Jean met Yvan Craipeau, an appealing, fiery young man who was the spokesman for a group of protestors who had come to the theatre to demonstrate and to distribute political literature. In their view, the current production at l'Atelier was slandering Lenin and presenting a distorted picture of the Russian revolution. They had come to apprise Charles Dullin, the respected director of the theater, of their opinions and, naturally, to gain some attention for themselves. After the *manifestation,* van Heijenoort approached Craipeau to find out more about the demonstrators and their goals, and a chord was struck. As he spoke with Craipeau, he immediately felt a bond. Within minutes, he was convinced they had the same way of looking at the world.

Although his Marxist opinions had not yet solidified, van Heijenoort's leanings were most definitely in the direction of Trotsky's theory of world revolution. The protestors all belonged to the *Ligue Communiste,* the local Trotskyite organization. To Jean, the specifics of the particular complaint were less important than their general tone. When Craipeau, a former student who had given up his studies to become a full-time revolutionary, invited him to come to some of the *Ligue's* discussions, he quickly accepted.

In 1932, the *Ligue Communiste* was a small band of no more than twenty deeply committed true-believers who felt no need for official membership cards. At their meetings, Jean's first impression of having found kindred spirits was strengthened. Even more than Gaby, these were people who spoke his language. They talked about political theory and current politics and Trotsky's ideas of the continuing revolution. He was impressed by Raymond Molinier, the leader of the *Ligue,* who, like Craipeau, had the answers he wanted to hear, and his complete conversion was lightning-quick. The seeker had found his ideological home. By the spring of 1932, he was considered a regular member of the group. When he could, Jean took part in the *Ligue's* activities, attending meetings and selling *La Verité,* their weekly newspaper, at the metro stations in the working class neighborhoods. As soon as it got dark, he and his friends would make the rounds, putting up posters—illegally, because they had no money to buy the required stamps. Often as not they got caught and would end up at the local police station.

A few months later, van Heijenoort and his comrades became totally involved with the issues of the German political scene, where Adolf Hitler was breathing fire and the Nazi party was making enormous gains in popular support. At that point, if Jean had not already crossed paths with Craipeau and his group at l'Atelier, he surely would have become engaged through some other means. As all of Europe followed the turbulent and frightening spectacle of the elections in Germany, Jean could no longer concentrate on mathematics. To his mind, a world crisis was in the making, and he had to join in some effort to stem the tide.

To the Trotskyites, the most disturbing aspect of the situation was that the German Communist party had grossly underestimated the power of the Nazi movement. They did not take Hitler seriously, and even further off the mark, the Party leadership declared several times that the Nazi party was on the verge of disintegration. While various voices on the left were urging the German Communists and Socialists to form a united front against the Nazis, Stalin's directive to the German Communists was to refuse such action. Incredible as it may now seem, the German Communists' stated position was that the Socialists were really "Social Fascists" and as such they were more dangerous than the Nazis. An official Communist publication declared that all the forces of the party must be thrown into the struggle against the Social Democrats. Thus, the two largest working class parties, the Social Democrats and the Communists, remained divided and inert while the Nazis grew ever more powerful.

In response to this paralysis and egregious misjudgment, Trotsky wrote a stream of articles and pamphlets which, according to many close readers, are the most brilliant short pieces of his exile period, including "What Next?," "The Only Road," and "What is National Socialism?" Unfortunately, at the time, only the converted benefited from his wisdom.

Describing the conflict between the Stalinists and the Trotskyites in France in his own memoir, *With Trotksy in Exile*, van Heijenoort wrote about a particular mass meeting called by the French Communist Party on July 27, 1932, at Bullier in Paris:

> Bullier was a large and popular dance-hall at the end of the Boulevard Saint-Michel, which could hold several thousand persons and was used from time to time for political meetings. The *Ligue* [the Trotskyite group] decided to be heard, to explain once

> more that the Socialist and Communist organizations should form a united front against Hitler. The hall was packed. Perhaps twenty of us stood in the midst of the crowd. After one or two speeches by official speakers of the Communist Party, who repeated that the main enemy in Germany was the Social Democratic party, we opened fire. Raymond Molinier [the leader of van Heijenoort's group] shouted, "We demand the floor for a five minute declaration!" He added a few words about the seriousness of the situation in Germany and the necessity of a united front against Hitler, but he did not get far. At a sign from Pierre Semard, a leader of the French Communist party who had made a specialty of persecuting the Trotskyites, the attendants, who had already taken positions around us, closed in, grabbed some chairs, and started to club us. I was one of the more seriously injured; my friends dragged me out with a bloody head.

As his bloody head healed, his conviction of being on the right side solidified. Meanwhile, the situation in Germany got worse and worse, and van Heijenoort's feeling of urgency became all-consuming. Although he did finish his second year at the Lycée Saint-Louis, it was not in the style to which he and everyone else had become accustomed. For the first time in his school career, there were no calls to the podium; not a single *prix d'excellence*! The fact that he was completely abandoning his studies and in effect renouncing fourteen years of dedication was even more shocking. Against all odds, he had arrived at the threshold of the École Normale Supérieure, the best university in France for the study of mathematics. In spite of the notoriously difficult entrance exams, there is little question that, as was his custom, van Heijenoort would have passed with honors and been granted a *bourse* for the *École Normale.* But he had come to the turning point—in fact he had already made the turn. Without reflection or hesitation, he was ready to make a leap into unknown territory and to serve a cause led by a man he had never seen.

Raymond Molinier had been watching van Heijenoort closely, ever since Yvan Craipeau had introduced the new comrade to the *Ligue's* discussion group. Deeply impressed by his talents, his fearlessness, and the intensity of his commitment, Molinier now proposed an assignment more risky but also much more interesting than selling *La Verité* and attending mass meetings. He asked Jean if he would go to Turkey to be Trotsky's secretary.

With his typical modest understatement, van Heijenoort later said, "Somebody was needed, and I suppose one of the reasons Molinier chose me was that I could read Russian." No doubt, the fact that he knew Latin, Greek, German, history, mathematics, and physics was not seen as a hindrance either. In any case, now, school was out in every sense and the former schoolboy was ready to prove himself a man, ready to take on the world. And what about his love of scholarly work? His talent in mathematics? A career? His mother? His response was, "How can you compare that with having a chance to change *everything*? I thought I was joining the staff of the World Revolution on its wings." Even in the volatile 1930s it was an extraordinary decision, but he made it with extraordinary equanimity. "I didn't think twice," he said when he recalled the actual moment of choice.

Certainly, there were other students at the *lycée* who were Marxists, who protested and held meetings and discussions and engaged in radical politics in response to the events in Europe in 1932. Most, however, were of a Stalinist rather than a "left oppositionist" bent. In any case, as far as Jean knew, no one among them felt inspired or compelled to give up everything for his ideals. So what made *him* do it?

At the moment of his choosing, it was the urgency of the times that made him decide to take drastic action: the political climate of France, Germany, Russia, and Europe in general, and of course, most particularly, the rise of Nazism in Germany. He saw the world heading towards Armageddon and he had to do *something*. He, as a moral, thinking person was not going to remain unengaged. He passionately wanted a righteous world and he was willing to do almost anything he thought would work to achieve that end. This feeling was what impelled his decision to join the Trotskyite group in Paris in the first place. Once he had made *that* commitment, becoming Trotsky's secretary was simply a next step and, in his frame of mind, an honor he could hardly refuse.

Taking a longer perspective, van Heijenoort in later years gave equal weight to other reasons: his father's death, the unjust circumstances that surrounded it, and his personal need for a father substitute were compelling emotional factors in his desire to serve the outcast Trotsky. And following from this, there was his entire life history within the general background of the historical period in which he had grown up. He had witnessed the killing and devastation of World War I, he had seen the cemeteries. He had experienced personal injustice and seen general inequality. He knew all about unfairness, exploitation and depression. In later years, van Heijenoort told one friend, "Any one of three or four reasons could have made me give up my studies in favor of politics, but I had them all. I was overprogrammed."

But part of the question remains; if van Heijenoort really wanted to change everything, what made him choose the exiled Trotsky's path rather than the mainstream of the Communist party movement as so many of his generation did? Why did he choose "the loser," one of his younger colleagues bluntly asked? Others put it more gently, but nevertheless it was a question he was frequently asked. After all, in 1932, the "mainstream" Stalin had not yet revealed much of the savage brutality that was later to become his trademark. If he had, the answer would have been much simpler.

One of the most cogent responses van Heijenoort gave to the "why Trotsky" question was in a letter (June 1982) to Jean-Yves Girard, a young colleague in Paris who, after reading van Heijenoort's account of his years with Trotsky in *With Trotsky in Exile*, was unsatisfied and wrote to van Heijenoort asking for more details. In his answer van Heijenoort said:

> As for my choice between T[Trotsky] and S[Stalin], I would almost say it was a question of style, of literary style, even. It was enough to read five sentences of one and five sentences of the other. Intellectual conviction also. I was a convinced and militant Trotskyite before going to Prinkipo, before thinking that I would go. So it wasn't personal contact that was at issue in

> the beginning. On a less rational level, the fact that S. was in power and T. in exile certainly was very important because that was connected to things which were deeply embedded within me (for example, sensitivity to unhappiness). I don't know how I would have reacted to a Trotsky in power. But, on the other hand, T. in Prinkipo didn't seem very much like a loser to me. Up until the point when Hitler came to power, his [Trotsky's] return to the Kremlin was not unthinkable. (Think of the Kirov affair!) No, the loser was S., who had abandoned Leninism, and I felt the extermination of all the old Bolsheviks during the purges as a sign of defeat for Stalin.
>
> Certainly, chance also played a part. I met Trotskyites when I was very young. Things might have happened differently. Or could they have, really?

To other people, at other times, and depending upon his listener's political sophistication, van Heijenoort would expand his list of reasons. He would cite Trotsky's ideas about the tempo of industrialization in Russia, his opinions on China and, later, on the war in Spain. He would enlarge upon the quality of Trotsky's writing and explain the charismatic effect of his great intellect, a quality that was recognized by his enemies as well as his supporters. Finally, in anticipation of, if not in direct answer to, the inevitable bottom-line question, he would reiterate, "In 1932, I didn't think I was picking a loser; I thought I was picking a winner, and the reason I thought we would win was that I knew we were right."

What a romantic ring that has, half a century later. Yet it is completely understandable. In the face of all the adversity van Heijenoort had known, and in spite of his early painful feelings of alienation, he had acquired the optimism that goes hand-in-hand with deep and passionate conviction. At twenty, he was hardly the first person to respond with a "right makes might" argument. To him the question was not, why would he go to Trotsky in Turkey; but rather, why *wouldn't* he?

# II

Chapter Four

# The Old Man on the Island of Princes

Turkey: 1932–1933

*Leon Trotsky*

His following consisted of tiny sects of young people, sincere, largely inexperienced, sometimes feckless. Personal tragedies, incalculable sufferings beset him, but he remained erect and combative, faithful to his vision in both its truth and error, its insight and blindness. Even those rejecting his every word must recognize that in the last ten or twelve years of his life Trotsky offered a towering example of what a man can be.

IRVING HOWE, *Trotsky*

IN OCTOBER 1932, Jean van Heijenoort, the renegade *taupin*, forsook abstract algebra, complex variables, the École Normale, his mother, his girl friend Gaby, and Paris itself. All anticipation, he set off for Turkey to join Leon Trotsky, the exiled Bolshevik leader, on Prinkipo Island in the Sea of Marmara. There, he would become a member of a household which included Trotsky, his wife Natalia Ivanovna Sedova, his six-year-old grandson Vsievolod (Sieva) Volkov, and a changing cadre of secretary-bodyguards. Van Heijenoort's work was going to be practical and applied, and only rarely would his highly developed and refined mathematical talents be called for. As far as he was concerned, that was just fine.

With money given to him by the *Ligue Communiste,* van Heijenoort had gone to Cook's in Paris and booked passage from Marseilles to Istanbul aboard the ship *Lamartine.* This early encounter with travel arrangements was the origin of one of van Heijenoort's later *idées fixes,* that Cook's was the best travel agency in the world, and for the next fifty years, he always bought his tickets there, in person, if it was at all possible.

Except for the brief period during the war when all the children of Creil had been sent to the country for safety, Jean had never been south of Paris. Even so, as he attended to all the necessary last minute details, he was not nervous about his departure or worried about how things would work out. He did, however, have the difficult task of explaining to his unhappy mother that he was "going far away," without being able to tell her precisely why, where, or for how long. Hélène, who by this time knew her son was deeply involved in politics, had to be satisfied with his promise to write.

By the time Jean boarded his train he was quite exhausted. Settling himself in a corner in the third-class carriage, he immediately fell asleep. At dawn, as the train was passing through the Massif Central, he awoke, and in the early light the mountains were blue. Never had he seen mountains so big and of such color, and he was overcome by their beauty.

The sensation of heightened awareness that attaches to important new experience possessed him. As his ship slipped away from the dock in Marseilles and he watched the receding land, it seemed to van Heijenoort that the land and sea had become a grand stage, set to render the drama of his passage to a new life. He slept on the deck, under the stars; he spent a day in Naples and a day in Piraeus, and he told his fellow passengers, who were mostly military people headed for the French colonies, that he was going to work for a publisher—which, in the broadest interpretation, was not false.

When the *Lamartine* docked in Istanbul, the eager recruit was more than ready for action. He dashed off the boat and onto the paddle-ferry for Prinkipo Island without seeing or being seen by Pierre Frank, the emissary Trotsky had sent to meet him. Two hours later, he arrived on the island, hired a horse-drawn carriage ("the horse wore a very nice red pom-pom", he would later recall) and rode in style to the villa where Trotsky was living.

Prinkipo, about twenty miles from Istanbul, is the largest island in a small archipelago in the sea of Marmara. In 1932, most of its year-round inhabitants were Greek fishermen, but during the summer the island was also a resort where wealthy Turks came to escape the oppressive city heat and enjoy the sea air. Many of these Turkish citizens had built solid and imposing villas for their vacations, and, by good fortune, a particularly attractive and large house was available for Trotsky, his wife Natalia, and his small staff to rent. It was right on the sea, in an ideal spot—"the most beautiful site in the world," was van Heijenoort's immediate impression. In this perfect place the young seeker found his hero, who, in his eyes, "was like a god."

Arriving at the house, van Heijenoort rang the bell and was met by the Turkish policemen on guard duty. Then Jan Frankel, the secretary he was to replace, came out to welcome him. Puzzled, Frankel asked, "But where is Pierre?" Realizing that Pierre was probably still looking for the new comrade at the port, Frankel brought Van inside the house, and as they stood talking, Trotsky came down the stairs to greet him. Van Heijenoort's first meeting with his leader lived up to his every expectation:

> He was dressed absolutely, completely in white, and he embraced me, in the Russian way, with three kisses on alternate cheeks. He was striking. His proud carriage, the set of his head with its halo of hair, his phenomenally lofty forehead, and his deep blue eyes with a gaze both powerful and sure of that power . . .

The glow with which van Heijenoort surrounded his leader kept him from seeing features others described as less than ideal. In *With Trotsky in Exile,* he quotes from the memoirs of Claire Sheridan (a British sculptor who had gone to Moscow in the twenties to do the heads of the Bolshevik leaders), and then argues with her assessment that Trotsky's nose looked as if it had been broken and that his pince-nez "rather spoiled an otherwise classical head." Van Heijenoort did, however, allow that Trotsky was one centimeter shorter than he was and that Trotsky's legs were not as strong as his "stalwart back." Although he suppressed his feelings at the time, there was one sour note in their first encounter. Upon seeing his new aide, Trotsky turned to Jan Frankel and said, "He looks like Otto." (Otto Schüssler was another of Trotsky's secretaries.) Van Heijenoort bristled but, of course, said nothing. He was hardly in a position to disagree with the great man at that point. Where all comrades were supposedly equal, how could he insist on being different? But forty-five years later he wrote, "Otto was blonde like me, but there the likeness ended for we were quite different in height and features." Oddly enough, a photograph taken at the time shows a quite remarkable resemblance.

Otherwise, everything else about van Heijenoort's new life in Prinkipo was good. On the island, van Heijenoort felt as though he had come to a new world. There was something special in the light and color. In the warm, salty air he had an immediate sensation of well-being: "I even breathed better," he

*Comrade Van (rear) in Prinkipo with Otto Schüssler and Pierre Frank. 1933.*

later recalled. To be sure, there was a strong emotional element at play, but there was also a very good physiological explanation for the way he felt. Since childhood, Jean had suffered from frequent respiratory problems, but he had not realized how adversely the damp, harsh weather of northern France affected him until he experienced the gentle Mediterranean climate. In that beautiful landscape he felt so well physically that nothing seemed beyond doing. The world had to be saved? Well, he, Jean van Heijenoort, was more than willing to take part in the struggle to overthrow both the evil capitalist system of the West and the pseudo-Marxist government that was leading the Soviet Union astray. Furthermore, there was no doubt in his mind that the effort would be successful. The young convert had come to Turkey, committed as any missionary of any faith, literally ready to lay down his own life, or to make any sacrifice asked of him. The striking man who embraced him and immediately dubbed him "Comrade Van" was the embodiment of everything that was right about the revolution. At that moment in his life, the newly baptized comrade experienced nothing but joy.

*Trotsky, Comrade Van (left front), and the other secretary-bodyguards, in the garden of the Prinkipo villa. 1933.*

With a minimum of ceremony, young Comrade Van was initiated into the routine of the household. He was shown his room, which he was to share with Jan Frankel. There were two narrow beds, two dictionaries, and a desk. He was shown the small arsenal of guns and rifles and shown the positions of the indoor guards. Because the Turkish policemen were thought to be easily susceptible to bribery, they were not trusted inside the house, and the real responsibility of protection lay with Trotsky's own bodyguards. Although Stalin's decision had been to exile Trotsky, exile did not mean safety. It was, therefore, the primary task of everyone living in the Prinkipo villa to do everything possible to prevent Stalin's agents or anyone else from doing injury.

Like all the other members of the group, the recruit was issued a revolver, a German Parabellum, and taught how to shoot it. For the next seven years, he never went anywhere without at least one, and more often two guns, worn on his body or kept next to him. Outdoors, his hand was always on the gun in his pocket and he learned how to shoot from that position. His second gun was carried in a holster, but when he slept, both guns were by his side. When his career as Trotsky's aide-de-camp was over and it was no longer necessary for him to pack a pistol, van Heijenoort said that for years it felt very strange to put his hand in his pocket and find nothing there.

Besides the crucial job of bodyguard, van Heijenoort's major responsibilities were to do translation and general secretarial work. After a brief training period, he replaced Jan Frankel as Trotsky's French secretary and Frankel returned to Paris as a leader of the *Ligue*. There was also a long and ever-expanding list of "minor" responsibilities and activities, some of them recreational, which kept the rookie comrade on the go every minute of his waking hours. He ran errands to Istanbul, went on regular very early-morning fishing and hunting expeditions with Trotsky and a Turkish fisherman, and did all manner of handyman work about the house and garden.

One of van Heijenoort's earliest and most unexpected assignments was a quick trip back to Paris with Sieva, Trotsky's grandson. The child had been left in his grandparents' care while his mother, Zinaida Volkov, deeply depressed, had gone

*Van and Sieva (Trotsky's grandson) on their way to Paris. Marseilles, 1932.*

to Berlin to seek psychiatric help. Van Heijenoort thought Sieva was "a sweet little boy and no trouble to anyone," but Trotsky, it appears, was angry with his daughter for leaving her child behind. So Van was asked to take Sieva to Paris, from where someone else brought him to his mother in Berlin. Like all of the Trotsky family history, the story would have a tragic ending. In January 1933, less than a month after she and her son were reunited, Zinaida committed suicide. Little Sieva was sent to Vienna to live with friends, then back to France to live with his uncle Liova, and, eventually, about a year before Trotsky's assassination, he was brought to Mexico to stay with his grandparents and his old friend, Comrade Van. Years later, in spite of the tragedy that had befallen him immediately after their trip together in 1933, Sieva would remember Van with enthusiasm and affection. "I liked him very much. He was always telling me jokes."

* * *

Lev Davidovich Bronstein, known to the world as Leon Trotsky and to his intimates as L.D. or "the Old Man," had been in Turkey for three years when young Jean van Heijenoort arrived on the scene. "When I first met Trotsky in Prinkipo," van Heijenoort wrote in *With Trotsky in Exile*, "he showed no outward sign of the series of ordeals that had brought him to this point." He was referring most directly to Stalin's persecution of Trotsky, but the "series of ordeals" had begun well before the revolution and before his conflict with Stalin. (The following mini-biography may be helpful for those who need a little of Trotsky's history.)

Lev Davidovich was born in 1879 in Yanovka, a village in the southern Ukraine, the youngest son of a prosperous but barely literate Jewish farmer and his more gentrified Jewish wife. Young Lev showed prodigious talent as a student, and therefore, when he was nine, his proud parents sent him to live with an older cousin in Odessa so that he might continue his education there. Odessa was a cosmopolitan city with a European cultural and intellectual climate that the farm boy found stimulating and exciting. The cousin, a publisher, became Lev's mentor, and through him Trotsky came to love everything about literature. Books, ideas, ideologies became the central thing in his life. "People passed through my mind like random shadows," he wrote in *My Life*, his autobiography. He was characterizing himself as a youth but with rare exceptions this remained true for the rest of his life. Ideas were more important to him than people.

Compared to van Heijenoort, Trotsky's personal circumstances were very comfortable—the Bronstein family had servants, whereas the Balagny women *were* servants. Nevertheless, through his reading, Lev became aware of and sensitive to the misery, cruelty, and brutality that existed all about him. Like many another serious child, he developed a strong sense of morality and gravitated toward utopian ideals. Describing the general picture, Irving Howe, in his excellent "small book," *Trotsky,* writes: "For a young man of independent spirit growing up in late nineteenth century Russia, rebellion was almost inevitable." (In the same vein, Jean van Heijenoort began what he called his "little book" with the personal statement, "I was in total revolt against society.")

In his last year of school, Lev Davidovich became a populist and then, in fairly short order, joined a group of activists with Marxist–socialist tendencies. At eighteen, in his first year at the university, he decided to become a professional revolutionary and a year later, in 1898, he was arrested for organizing and being part of the Southern Russia Workers Union and participating in anti-czarist political activity. For this crime he was sent to prison in Odessa for almost two years and then sentenced to exile in Siberia for four years. In 1900, in the Moscow transfer prison, Trotsky married Alexandra Lvovna Sokolovskaya, one of his closest comrades, who had been arrested and imprisoned along with him. Following the surprisingly humane Russian custom, the newlyweds were given permission to go to Siberia together.

Because political exiles in Siberia enjoyed a certain amount of freedom and mobility, Bronstein found work there as a journalist. He wrote essays and newspaper articles, and he also had plenty of time to read, think, and formulate his political ideas. He and Alexandra had two daughters in quick succession, and they became part of the community of dissidents. As revolutionary activity increased in Russia, various leftist organizations surfaced even in Siberia, and literature was smuggled to these groups. One day "a number of books printed on extremely fine paper" (in his autobiography, Trotsky, the bibliophile, could not resist remarking on this fact) arrived from abroad, among them Lenin's pamphlet, *What Is To Be Done?* The text addressed the need for "a centralized organization of professional revolutionaries who would be bound together by the iron discipline of action," and when it was circulated, the reaction was explosive. According to Trotsky, there was an epidemic of escapes by exiles no longer willing to be confined, and after two years in Siberia, Lev Davidovich Bronstein responded to Lenin's clarion call with the same decision: "I had to escape from exile," he wrote. "My handwritten articles for the *Siberian Union* immediately looked small and provincial to me in the face of the enormous task which confronted us. I had to look for another field of activity."

Alexandra Lvovna agreed and, according to Trotsky, insisted that he must go. Lenin's pamphlet had arrived in summer and before autumn he was gone, leaving his wife and two babies behind. She helped in the preparation and concealed his departure for days afterwards. First by cart, hidden in a pile of hay, and then by train, he made his way to western Russia, aided by a network of other revolutionaries. His friends provided him with a false passport into which, with inspired audacity, he inscribed "Trotsky," the name of the chief jailer in Odessa.

"I wrote it in at random," he said in his autobiography, "without even imagining that it would become my name for the rest of my life." He also noted that as he left Siberia he had, as reading material, a copy of the *Iliad* in the Russian hexameter of Gnyeditch.

From Russia, the newly christened Trotsky went on to Vienna, Zurich, Paris, and, in the fall of 1902, London. There he arrived at Lenin's lodgings in the middle of the night. Banging on the door and waking the household, he is alleged to have shouted from the street, "*Peró* [the pen] has arrived." He was then twenty-three and very much more a man of the world than the young van Heijenoort would be when he arrived in Turkey thirty years later. Trotsky had a wife and two children, he had been in prison, he had traveled through Europe and seen a bit of the world. Confident as a writer, he soon established himself as a first-class orator. Lenin and the cadre of revolutionaries surrounding him were much impressed with his intellectual abilities, his conviction and his energy. Almost immediately he became part of the inner circle of revolutionaries, and four months after his arrival in London, Lenin proposed him as a member of the editorial board of *Iskra*, the Marxist journal.

Trotsky's next assignment was a lecture tour on the continent. This time in Paris he met Natalia Ivanovna Sedova, a student member of the revolutionary movement. She was his guide, his helper, and soon his lover. When Trotsky returned to Russia in

1905, Natalia came with him. Although Trotsky and Alexandra Sokolovskaya were never officially divorced, once he left Siberia, in effect their marriage ended. Nevertheless they remained friends, and during the periods when Trotsky returned to Russia, he occasionally saw her and their daughters. In spite of her separation from Trotsky, the loyal Alexandra continued to suffer exile and persecution on his account, first by the czar and then by Stalin. Meanwhile, Natalia Sedova became known as Trotsky's wife and remained with him throughout his exile, until his death.

Lenin was recognized as the leader of the Russian revolutionary movement in exile, respected and admired by his followers, but by 1903 Trotsky, among others, found himself in basic disagreement with Lenin's views on party organization. At the Second Congress of the Russian Social-Democratic Labor Party in London there was a factional split which resulted in the formation of two groups: the majority, led by Lenin, and the minority led by Martov. Trotsky joined with Martov who held that the party should be open not just to professional revolutionaries but to anyone who believed in their goals.

Although Trotsky had aligned himself with Martov in 1903, he often remained independent and tried to find ways to reconcile the factions. For at least a decade there were many tactical if not ideological differences between Trotsky and Lenin, and it was not until the October revolution of 1917 that Trotsky would join Lenin's party and assume his pivotal role.

In 1905, Trotsky returned to Russia and became one of the leaders of the failed 1905 revolution. Again, he was arrested and this time, banished to Siberia for life. Again, he escaped, this time accompanied by Natalia and their infant son, to Vienna. There he remained, working as a journalist until the beginning of the First World War. Then he and his family, which by now included two sons, fled to France, Spain, and finally to New York City where he worked as a printer in a shop on St. Marks Place for a few months. His early love and knowledge of printing, learned from the cousin in Odessa, stood him in good stead. After the overthrow of Czar Nicholas, in May 1917, Trotsky, with some difficulty, succeeded in returning to Russia to join with Lenin, who also had only just returned from *his* years of exile in Switzerland. Ironically, the Germans had permitted Lenin's train to pass in the hope that the Bolsheviks would create dissent and undermine the Russian war effort. Once again on Lenin's side as a Bolshevik during this period of shifting alliances and conflict among the various factions of the Socialist movement, Trotsky was arrested by the Kerensky government.

Finally, came the revolution—the *big* one—and Trotsky, who by then had been released from prison, was ready to play his crucial role. As head of the Red Army, which Lenin had asked him to form, he staged an armed coup d'état in Petrograd. The government buildings and the Winter Palace were taken, and, almost miraculously, the grand plans of what had been a rather small Bolshevik party were realized. Under the leadership of Lenin, with Stalin, Trotsky, Bukharin, Zinoviev and others in the cabinet, the Bolsheviks took over. Even to these leaders and planners, it was a revolution that at that point seemed improvised rather than historically inevitable—a fact to bear in mind when considering how realistic it might have been in later years for the exiled Trotsky and his followers to think they had a chance at a return to power.

As long as Lenin was the head of government, Trotsky remained in a favored and powerful position in the Politburo, as minister of foreign affairs and as commissar of war. After Lenin's death in 1924 and Stalin's succession, the picture changed drastically and a new "series of ordeals" began. The ideological disagreements that had always separated Stalin and Trotsky while they had both been members of Lenin's cabinet became increasingly intense and personal. Trotsky advocated a program of continuing or world revolution while Stalin held to the notion of socialism in *one* country—Russia—first; only when the time was ripe did he think it should be attempted in other countries. Trotsky's followers called this the "bureaucratic view."

*"Le pouvoir ne se partage pas"* (power cannot be divided) was the beginning of van Heijenoort's assessment of Stalin's motivation for his ruthless and relentless plots and actions against Trotsky as well as his other former allies. As soon as Stalin established his control of the party machine, he methodically began carrying out his decision not only not to share leadership

but also not to tolerate divergent opinions of any kind. Trotsky had become the acknowledged leader of what was known as the "Left Opposition." He was a popular man: a successful military leader, a brilliant writer, and an eloquent speaker. Stalin feared all of these attributes in his rival. His solution was to eliminate Trotsky from the corridors of power, to effectively squelch all dissident ideas, and any possible further challenge to his own leadership.

Stalin dismissed Trotsky as commissar of war in 1925, expelled him from the Politburo in 1926 and from the party in 1927, and exiled him to Alma Ata, in Turkestan, in 1928. A year later, Stalin began to worry, perhaps quite rightly, that two thousand miles from Moscow was not far enough, and he ordered Trotsky and his wife Natalia to leave the USSR. Their sons, Liova and Sergei, now in their twenties, were given the choice of leaving or remaining in Russia. Liova, the elder, went with his parents, but Sergei, who was in the midst of his studies, stayed, only to become yet another victim of Stalin's pathological cruelty; for although Sergei abstained from politics, he was nevertheless persecuted and eventually shot in the wake of the Moscow Trials of the 1930s.

Van Heijenoort's speculation was that as far back as 1929 Stalin considered having Trotsky assassinated, and would gladly have had the job done, but decided that the risk was too great because the repercussions would be unpredictable. The former commissar of war still had friends with power and influence who might make use of such a drastic act to turn the tables against Stalin. Instead, Stalin banished him to Turkey, where he would be isolated and without financial resources. Under these conditions, Stalin figured he could keep his enemy at bay. But whatever Stalin's expectations might have been, Trotsky in Turkey did not close up shop. He was, after all, a man of great imagination, energy, and drive, and well acquainted with adversity. He had already spent twenty years in prison or exile and, if nothing else, he knew how to put his time and effort to good use. Tenacious to the end, giving up was out of the question for Lev Davidovich.

In the passport issued to Trotsky upon his expulsion from the Soviet Union, his occupation was listed as "writer." The description as given was not false and, after all, what else could have filled the space? In *My Life*, Trotsky wrote,

> The very word 'author' sounded to me as if it was uttered from some unattainable height ... From early years my love for words had now been losing now gaining in force, but generally putting down ever firmer roots. In my eyes, authors, journalists, and artists always stood for a world which was more attractive than any other, one open only to the elect.

Thus in Prinkipo, deprived of his sword, the former commissar of war picked up the pen for which he was justly famous and sent his "never ending stream of letters, pamphlets, articles, and tracts" to individuals, political organizations, and publishers all over the world, always, in one form or another, attempting to convince people that his views of world history and politics were the correct ones. Most of the articles were written for and distributed by the Trotskyist press, but from time to time, in order to make money, he wrote for "bourgeois" journals in the United States and even in Germany, before Hitler's takeover. Also, he worked on his remarkable autobiography, his histories of the revolution, and his biographies of Lenin and Stalin. Some of Trotsky's best work was written during this period. This was the literature that impressed Jean van Heijenoort so much that he was willing to follow the man who expressed his ideas with such vitality and grace.

* * *

Every morning after breakfast, Maria Ilinishna Pevzner, a Russian secretary, used to come from Istanbul to the villa in Prinkipo. Trotsky would often spend several hours dictating to her, walking back and forth between his study and her office. From his adjacent room van Heijenoort could hear them at work. "I could hear his well-articulated sentences, rhythmical and melodious. One could guess what the power of this voice had been when addressing a crowd, at a time when the art of oratory was not yet aided by electronic devices." One can also imagine the effect this had on Van, the worshipping neophyte.

When composing letters in languages other than Russian, Trotsky sat at his desk to dictate. This was the method he favored when he worked with Comrade Van, who was responsible for everything written in French, and for translating articles written in Russian into correct French. With his excellent background in Latin, Greek, and German, and natural talent for languages, Van had taught himself the Russian alphabet and, using a dictionary, he could read and write reasonably well.

The new secretary was also struck by the precise manner of Trotsky's speech:

> Whether he spoke in Russian or a foreign language, his lips shaped the words clearly and distinctly. He was always irritated at others' confused and hasty speech and forced himself to enunciate with precision. In conversation with visitors in his study, his hands, at first resting on the edge of the work table, would soon begin to move with wide, firm gestures, as though aiding his lips in molding the expression of this thought.

"Because I knew Russian," was one of the reasons van Heijenoort later gave for his having been chosen to be part of the intimate circle that surrounded Trotsky. He was too modest, or thought it bad form, to add that he was extremely attractive to the organization in a number of other ways: brainy, dedicated, disciplined, thorough, and willing to do almost anything. Living among people of explosive temperament, it also helped that Comrade Van was quiet and well-mannered, and—unknown to Raymond Molinier, who chose him in Paris—he possessed skills that came to be highly regarded in Prinkipo. Serendipitously, the Trotskyites had found a gold mine.

It was not too long before Van's myriad talents were discovered and exploited. Trotsky, whose wit had a teasing, sarcastic bite, gave him a new name for each revelation. After Comrade Van repaired the water pump, he baptized him "Comrade Technocrat." A few months later, Comrade Van succeeded in obtaining the precious visas for France at the French consulate in Istanbul, the documents which made it possible for Trotsky and Natalia to leave Turkey and establish residence in France. At that point, "Comrade Technocrat" was elevated to the position of "minister of foreign affairs." Later still, in Mexico, Trotsky took to calling his most faithful secretary, "*Uzhe*," the Russian word for "already," because by then Van had come to know him so well he could anticipate his needs and wishes. Trotsky would ask him to bring something or do something and *Uzhe*'s response would be, "I've done it already." The name Trotsky did *not* use was 'Jean'. From 1932 on, only Jean's family and his most intimate friends continued to address him by his first given name.

Satiric, but also affectionate, Trotsky's titles were signs of the master's acknowledgment of Van's expertise; but there was yet another name the acolyte had to endure at every meal: "Comrade Molokan," Trotsky would ring out in his big voice, because instead of coffee or tea, Van drank milk (*moloka* in Russian). The joke was that the Molokans, an ascetic religious sect founded by the Doukhobors, drank milk on fast days, contrary to customary Orthodox observance. With that title, Comrade Van, the milk drinker, was given the status of a religious fanatic.

Van responded to it all with good cheer. Then and always he had the virtue and ability to laugh at himself, and since at that time, his admiration for Trotsky was unbounded, he felt flattered by the attention rather than perturbed by the irony. "That man," as van Heijenoort often referred to him, could do no wrong. His leader was as near to perfection as a human could be: noble, wise, just, and fearless. Like everyone else in the group, Comrade Van consciously copied Trotsky's ways of thinking and speaking, and deliberately took on his attitudes and mannerisms. As a sign of deep respect, he always called him Lev Davidovich, finding the Russian patronymic perfectly suited to convey the mixture of admiration and affection he felt. Only years later would he refer to him, in the familiar fashion of the other comrades, as "the Old Man."

With hindsight, van Heijenoort would come to change his perceptions in many ways; sometimes his memory of the quality of an event shifted with his mood on the day he happened to be recalling it, but regarding his early feelings for Trotsky, his description was singular and constant: "I was in love." Nothing less would do to convey the intensity of the emotion he carried for the ideas and the man who embodied them. Part and parcel, he loved them both, and because they were inseparable, it was impossible to be critical of either.

Decades after the fact, whenever van Heijenoort was asked to account for his early career with Trotsky, when such an attachment was long past being easy to understand, he always began with a carefully laid out exposition of the historical facts which led him to his choice: the terrible effects of World War I and the postwar trauma in France; the economic depression; the situation in Germany; and so forth. Because he was an intellectual, he emphasized the intellectual over the emotional; it was his intellectual conviction, his commitment to the idea and the ideal of Marxism, his total dedication to the Trotskyist cause, with heavy emphasis on the "total," that was most important to him. It was a cold, hard job of revolutionary work that van Heijenoort was ready and eager to embrace. "If you don't understand that, then the whole thing doesn't make sense," he would say, "the whole thing" being how and why he traded his life as a scholar for that of a professional revolutionary.

In other words, to him, his choice was logical and reasonable. Except for his relationships with women and the mysteries of romantic love where, like a storybook Frenchman, he would throw up his hands, look to heaven, and without apology say "It's beyond me," van Heijenoort's later explanations about his beliefs and actions were always based upon principles of logic and reason. There was a complex set of intellectual arguments for why he had done this or that. But lurking just below the surface of those arguments were the personal needs, the drives and passions, that inspire action. Van Heijenoort knew that about himself as well or better than the next person. A close and constant reader of Blaise Pascal's *Pensées*, he understood and identified deeply with its insights even though he knew Pascal was thinking of another kind of belief. Comrade Van knew all about the "heart's reasons that reason knows nothing about," though for many years he did not speak readily about that side of the story. However, when van Heijenoort did finally come to talk about the personal forces which shaped his choice, he never omitted the statement: "I believe in strong feelings. Without that, there is nothing."

So, van Heijenoort allied himself with Trotsky because intellectually he agreed with him. He believed in international Marxism and the principles of permanent revolution as a solution to the world's problems, and he was opposed to the limited, "bureaucratic" Marxism Stalin had established in the Soviet Union. Because of Stalin's decision not to oppose Hitler's rise in Germany and to refuse a Communist-Socialist alliance against Hitler, Van, along with many others, felt a desperate need to take action of some kind. Therefore he joined Leon Trotsky, the Marxist who understood better than anyone else what was happening in Germany and was trying with all the admittedly small means at his disposal to do something about it.

But of course there was more to it: as he embodied the tragic and the heroic, Trotsky was irresistibly appealing to the young van Heijenoort. In the most simplistic terms, in Lev Davidovich he had found, at last, someone he could cast in the role of leader, mentor, and father. Trotsky was even the same age that Van's father would have been. And then there was the injustice of Trotsky's situation: he was an outcast, a figure Van could identify with on the deepest level because it was part of his own heritage. Van didn't have to learn about suffering—he had had those lessons early.

There was also a very positive side to the attraction. Trotsky's vivacity in the face of adversity was inspiring. At fifty-three, this "father" was a man of prodigious gifts and boundless energy who went about his business of enlightening his comrades, fellow travelers, and whomever else he could persuade to listen. Even in exile, stripped of his former rank, the Old Man had the air, if not the actuality, of power and authority that seemed to give him at least some control over the events that governed his life—or at any rate, he appeared that way to young van Heijenoort. On Prinkipo, the Princes' island, in Turkey the new disciple discovered a chance for a kind of personal interaction and total involvement, a way of dedicating himself to an idealized person and an ideological cause, that he had never experienced. Certainly the risks were considerable but so were the rewards, because as Trotsky's protector and helper van Heijenoort himself would acquire a sense of power. The pervasive helplessness he had experienced as a child, beginning with his natural father's death and the haunting memory of no one's knowing what to do, and continuing with a series of events over which he had no control, would not be undone, but now, at twenty, he felt he could take an active part in his destiny and, in exalted moments, even shape the destiny of others.

Perhaps unfortunately, the inner needs of the two men did not quite coincide. Trotsky needed Van as secretary and bodyguard, but not as a son. He already had two sons, one of whom was an integral part of the inner group, and although his relations with Liova were always tense, Lev Davidovich was not looking for a replacement. His appreciation of Van's contributions was great, and he applauded the young comrade's myriad talents; but, as Van himself would come to realize, Trotsky had no deep interest in him as an individual. The deep love Van felt for his leader was not reciprocated or even understood in any meaningful degree. By his own choice, Van had given up literally everything to join in a risky and unpopular movement, yet the leader of that movement had never asked him, except in the most general way, about what had propelled him. Trotsky expressed no interest in Van's family or his student days or anything else outside the political scene. It was not that he cared less about Van than others, but rather that he did not care or think about people's feelings. With rare exception, Comrade Trotsky was not concerned with the hows and whys of any individual except as he or she contributed to the work of the movement. At first this lack of interest did not matter to the devoted secretary because he was so taken up with the larger purposes of his new life. Not until the exiles were in Mexico did Van begin to see a serious flaw in Trotsky's insensitivity to the feelings and opinions of others—even those closest to him.

But for the time being, at the villa in Prinkipo, the honeymoon was in full flower. One of Van's first assignments was to retranslate an old Trotsky article, to remove the overly verbose flourishes of the previous translator. His self-taught Russian was more than adequate for the job. He passed the "test" and was graduated to work on the articles currently being written. Guard duty was a second test the new comrade was able to pass, although he said later that it was physically one of the hardest things he ever had to do. The hard part was not handling a weapon; he had arrived knowing how to do that. As a boy in Creil, he had participated in the popular Sunday sport of shooting at beer bottles. With good eyes and a steady hand, he was a good shot. It was the lack of sleep that van Heijenoort thought was physically painful. "To have somebody come and shake you at two in the morning, and you know that you have to get up because it is your turn ... that was the worst."

And how realistic was all that guarding? What could three or four men do against the long, powerful arm of an enemy like Stalin? Fifty years later, van Heijenoort's answer was,

> I never deluded myself that we could stop a determined and organized attack on Trotsky's life. If the Soviet secret police had given the order, how could we stop them? No, we were just doing what we could, thinking we might at least prevent an attempt by an unbalanced person or by one of the many White Russian emigrés who had settled in Istanbul after the Revolution.

In all the years he lived with and was close to Trotsky, van Heijenoort would almost never be frightened or worried about his own safety. Alert, yes; tense, yes; ready to shoot if necessary, ready to come between his leader and an attacker, yes; always aware that his life was at risk as well as Trotsky's; but just plain scared, no. Guns, target practice, a certain amount of cloak-and-dagger were all part of the routine of life with Trotsky. And Van, who was twenty, a revolutionary, and a romantic, liked it.

* * *

Knowing, as we now do, what happened to Trotsky and his movement as well as to the Marxist revolution in Eastern Europe, it may seem like sheer naiveté or delusion that anyone could have been optimistic about the possibilities of Trotsky's return to power. A dozen devotees on a tiny island in the Sea of Marmara guarding the Old Man, typing his letters, translating his manuscripts—how could a logical person like van Heijenoort have been part of that coterie? Yet within a year, the group had succeeded in moving to France and had settled near Paris.

*Comrade Van, with gun, on the balcony of the Prinkipo villa. 1933.*

Twice before, Trotsky had returned from the obscurity of exile to become a leader of revolution in Russia. The second time, scarcely more than a decade past, he and his comrades had succeeded in changing the course of history. Trotsky himself had been the organizer and commander-in-chief of the Red

Army. Those had been violent, chaotic days, but the situation in Europe and particulary in Germany just after Hitler's accession to power was no less volatile. Someone might assassinate Stalin; revolution could take place in France or Germany. These were the dreams the Trotskyites hoped to activate.

Chapter Five

# On the Move

France: 1933–1935

*Van Heijenoort and Max Shachtman, at the French embassy in Istanbul. Van is clutching the French visas he has just obtained for Trotsky and Natalia. 1933.*

People nowadays think of the "thirties" as a block. They think that the political scene was fixed, but that wasn't true at all. The situation in Germany and France was changing all the time, and, then, there was the war in Spain ... Everything was very unstable. You never knew when there might be a chance to step in.

JEAN VAN HEIJENOORT, *1983*

A LONG-AWAITED SHIFT in the political wind enabled Leon Trotsky and his small band of followers to reach the desired shores of France. In January 1933, Edouard Daladier, a Radical Socialist, was named premier, and as an indirect result of this election, the prospects of the inner circle in Prinkipo improved dramatically. Less than a year after van Heijenoort left Paris without any notion of when or if he would return, he was back in France, still on the job as Leon Trotsky's secretary and general factotum, but now in much more familiar territory.

In early July 1933, as Trotsky's emissary, Comrade Van had gone to the French consulate in Istanbul, where he succeeded in getting the necessary French entry visas officially stamped into Natalia and Leon Trotsky's Turkish passports. Keeping in mind the workings of French bureaucracy, this was no small feat for the newly designated "minister of foreign affairs." However, prior to Van's final bit of legwork, the most important negotiation had already been done in Paris by advocates more experienced and sophisticated than he.

It was well known among the comrades that while the Radical Socialist party leaders might not be directly sympathetic to Trotsky's political views, they were not hostile to him personally. Aware of this new mood, Maurice Parijanine, one of Trotsky's translators and a man of some influence, aided by Alexis Léger, the much admired diplomat–poet whose pseudonym was Saint-John Perse, had actively lobbied on Trotsky's behalf. Raymond Molinier, van Heijenoort's first mentor, and his brother Henri were also involved because they, too, had access to important members of Daladier's party. The result was that these friends were effective in persuading the new administration to rescind the old expulsion order issued against Trotsky in 1916, and to grant permission for the exiled leader and his wife to enter the country and take up residence for an indefinite period.

If ever the group had reason to be encouraged, this certainly was the moment. To exchange the isolation of an island in Turkey for the ferment of France was a giant step in the right direction. They would be in a major center of political activity during what they knew was a crucial and volatile period in European history. They would have the potential for contact and action that had been out of the question in Turkey. And even if Trotsky could not act *directly* in France, at the very least he would have the opportunity to be a catalyst, to stimulate others to work in the interest of his ideal of international Marxism. With the right coalition of forces, the right conditions, the right leader (Trotsky, of course), the achievement of their goal did not seem completely out of reach to Van and his comrades.

Five days after van Heijenoort brought the passports from Istanbul, they were ready to leave. Acting as quickly as possible, no doubt for fear that Daladier might change his mind, they packed the archives and the books—most of which Trotsky had brought with him from Russia—in large wooden crates, and loaded them onto a barge which took them to directly to the *Bulgaria,* an Italian ship bound from Istanbul to Marseilles. Trotsky and Natalia, Comrade Van and the rest of the Prinkipo secretary-bodyguards followed in a launch. Meanwhile, Jan Frankel, the secretary whom Van had replaced, came back from Paris to Prinkipo to settle the affairs of the house and sell the boats and other miscellaneous belongings.

The group sailed out of Istanbul harbor and into the Sea of Marmara on the evening of July 17, one week shy of van Heijenoort's twenty-first birthday. Whatever personal regrets he may have had about leaving the island he had called "the most beautiful spot in the world" were overshadowed by the exhilaration he felt as he headed towards the greener political fields of France.

A week later, the *Bulgaria* docked in Marseilles, but to the waiting reporters' surprise, Trotsky and his wife were not on board. Shortly before the scheduled arrival, their son, Liova, had come alongside with a small boat and spirited them off to Cassis, while the rest of the party sailed into port. This

*France at last, shortly after their arrival in Saint-Palais. Clockwise: Jean van Heijenoort, Rudolf Klement, Trotsky, Yvan Craipeau, Jeanne Martin, Sara Jacobs. Saint-Palais, 1933.*

unexpected maneuver, designed by Liova to throw the press and possible enemies off the scent, worked beautifully—indeed so well that it caught the rest of the entourage by surprise and caused some difficulty for Comrade Van, who had to explain to the French customs officer why he and others in their party were carrying guns. Without Trotsky aboard, his explanation that he and the other men were bodyguards bearing arms to protect Trotsky was, understandably, not very convincing. The guns were confiscated, but otherwise the move was a success and spirits were high.

With careful planning and the cooperation of the French authorities, Trotsky and his staff managed to travel through France incognito. For a few months that summer they rented the

Villa Sea Spray, a large airy house near the resort town of Saint-Palais, on the Atlantic coast, and kept their place of residence secret from the press and, therefore, the public at large. In fact, although their motivations were different, during the entire period of Trotsky's exile in France, his followers and the French government consistently agreed on the need for secrecy. The Trotskyites were worried about assassination attempts, whereas the government was worried about political repercussions.

Even the owner of the villa in Saint-Palais, where Trotsky and his group lived for several months, had no idea who his tenants were until long after they had gone. Trotsky played his part in this game of hide-and-seek very well, spending nearly all his time indoors or on the grounds surrounding the house. Normally, in spite of the risk, the Old Man, like many leaders, resisted being reined in, but in this instance there were compensations that kept Trotsky from venturing farther afield.

Amidst the great secrecy about Trotsky's whereabouts, a select group of people, not only inner-circle Trotskyites but sympathizers and interested parties of various political persuasions, were invited to come to Saint-Palais to meet with him. They came, as individuals or as representatives of groups, to offer their views and to hear what the famous Lev Davidovich had to say about the current political situation, the Soviet Union, and the position of the "left opposition" regarding the official Communist party.

After the years of isolation in Prinkipo, where almost all of the comrades' activities had been focused on the exchange of written words, these were days of great excitement, of intense, face-to-face discussion of strategies for action as well as the unremitting definition and redefinition of Marxist doctrine and its proper application.

To facilitate all these meetings and exchanges, Comrade Van and the other secretaries and assistants shuttled back and forth to Paris, carrying messages, fixing appointments, picking people up at train stations and transporting them to the villa. Van Heijenoort, who kept track of such things, noted that in the three months in Saint-Palais, "some forty-five visitors, many of them foreigners, came to see Trotsky for political discussions."

André Malraux was among the first visitors to Sea Spray. He and Trotsky spent hours talking about a broad range of subjects: art in Russia after the revolution, the problem of individualism and communism, the causes of the Red Army's defeat in Poland in 1920, and the writer Céline, whom Malraux knew. Malraux later published an account of these discussions which van Heijenoort, ever a stickler for accuracy, would find annoyingly inaccurate in several instances. Malraux placed his meeting with Trotsky in the wrong year; he got Trotsky's age wrong by six years; worse yet, he described him as "gravely ill." While van Heijenoort would acknowledge that the Old Man had recurring back problems, or "lumbago," as well as occasional unexplained fevers, these episodes rarely kept him from his work and did not derail him from the course he had set.

Since 1923, Trotsky had criticized Stalin and other leaders of the Communist movement for their deviation from what he considered the true revolutionary goal. Even after he was exiled by Stalin, his goal was to put the revolution back on track and not to set up a rival party. But after Hitler's ascension to power in January 1933, when the German Social Democrats and the Communists, under orders from Stalin, refused to unite, Trotsky came to the conclusion that it was fruitless to continue to work for reform within the established Communist parties. In Saint-Palais, he and his group decided once and for all to abandon their critical position of left opposition and to create a new and independent international Marxist party, to be named the Fourth International. Under this banner, they planned to organize a cadre of leaders and to attract the other independent socialist and communist groups in Europe. As reported by van Heijenoort, one evening after a long discussion of plans, hopes, strategies, and new directions, one of the visiting French comrades said, "In short, what you are proposing is to begin all over again?"

"Exactly," Trotsky answered.

Trotsky was confident; his charisma and energy as a leader were undeniable, and his optimism suffused everyone in his group, including van Heijenoort. He was ready for a fresh start; large undertakings did not intimidate him, and in the full flush of enthusiasm he thought everything was possible. In those first months in France everything had indeed gone extremely well. There had been no difficulties with the local authorities in Saint-Palais or with the Sureté Générale, the national police. No restrictions had been placed on Trotsky's visa except that his place of residence had to be approved, and, since the police had been unexpectedly cooperative, the location of the "approved residence" continued to be unknown to all but the selected invited visitors.

Even in later years, van Heijenoort insisted upon his analysis that in 1933 and early 1934 the potential for the Fourth International to become a leading force in another revolution was there, and not just a far-fetched dream. The program of the newly created Fourth International was being accepted the world over by Marxists who were opposed to Stalin. They had only to seek out and seize the opportune moment to increase their strength and then they would return to power in the Soviet Union. Or if not, they would establish a base elsewhere from which to begin the drive toward world revolution. France or Spain or Germany might provide a foot- or at least a toe-hold.

Whether or not this argument has merit, there was a precedent for it: the Russian revolution of 1917 had been brought about by a very unlikely group of people—Trotsky among them—operating in an almost improvised fashion. In a chaotic time there had been a failure of leadership by the Mensheviks, and the Bolsheviks, led by Lenin, had seized the opportunity and taken control. If this changeover had happened once, why not again? If Trotsky had returned from exile twice, to participate in the 1905 uprising and to lead the army in the successful October revolution of 1917, was not a third, charmed return within the realm of possibility?

Sea Spray Villa had provided an ideal first stopping place for the exiles, especially in summer, but as fall approached, the weather turned cold and stormy. The Atlantic coast, which had been filled with vacationers, became deserted. Those who remained were conspicuous, a condition the Trotskyites were bent on avoiding. It was time to relocate, and again they were lucky in finding what seemed to be an ideal place.

*The Villa Ker Monique in Barbizon. 1933.*

Raymond Molinier had found the house in Saint-Palais and now his brother Henri came up with a secluded villa in Barbizon, a small town near Fontainebleau forest, where the painters Rousseau, Corot, and Millet had lived and worked in the mid-nineteenth century. Henri was able, aggressive, and resourceful; he knew everyone in and out of government, and unlike most of the comrades, he also had business contacts. (The Moliniers had a "business" of their own: an agency called "The French Collection Institute." The reports about this "institute" were far from favorable; it seems that the brothers used some of the most traditional methods for "collecting"—including threats and blackmail.)

By pointing out how well things had gone in Saint-Palais, Henri persuaded key officials in Daladier's administration to grant permission for the move to Barbizon, which was only thirty miles from Paris. Given a choice, Trotsky would have preferred to be in Paris itself, in the dead center of influence, but, from everyone else's point of view, there was too much risk. Molinier did not want to jeopardize his success thus far by proposing the capital city as a place of residence, and van Heijenoort worried about physical danger to Lev Davidovich. In any case, Van was certain that the French authorities would not have permitted it for fear that Trotsky and his comrades might take too great a part in daily political activities. So the group contented itself with the Villa Ker Monique in Barbizon and soon discovered that it was easy enough for Trotsky to get to Paris on a regular basis for meetings and talks. As in Saint-Palais, the owner of the villa did not know the identity of his illustrious tenant, but this time the secret went deeper yet, for neither the mayor of Barbizon nor the local chief of police were informed. In November, 1933, when Trotsky and Natalia moved in, only a handful of Trotskyites and a few of Daladier's Radical Socialists were aware of it.

Between their departure from Saint-Palais and settling into the villa in Barbizon, Trotsky and Natalia went to the Pyrenees for three weeks for a much needed period of rest and relaxation. As a result, van Heijenoort had a rare episode of freedom, a sort of trickle-down vacation of his own, with no one giving him instructions about where to go or what to do. He went to Paris, straight to Gabrielle Brausch, the young woman he had met at the theater when he was still at the Lycée Saint-Louis. He had kept in touch with her by letter from Prinkipo, and he had already been in brief contact with her since his return, but there had been little time for anything to develop. Now there was time, and the love affair was renewed. Again they walked along the Seine, talking about what had happened since they had last seen one another. Their conversations were about politics and love. Jean, the scholarly *taupin*, had changed most; known now as Comrade Van, he had found his vocation and was bent on convincing the world where salvation lay. Gaby, herself politically active, if not already convinced, was a willing convert. Almost immediately, they became lovers and were so deeply serious about one another that when Trotsky and Natalia returned, ready to move into the Barbizon house, Comrade Van felt he had to tell his "father" about the liaison.

The Old Man raised no objection. On the contrary, he was pleased and for good reason. Gabrielle Brausch was of the right political persuasion. Lively, energetic and willing to work, she seemed to be an ideal companion for Comrade Van, and not incidentally, she could be a valuable addition to the Trotsky staff. Gaby was asked to be an assistant to Natalia Sedova, a glamorless job with no salary (no one in the movement was paid), and she accepted. She gave up her position as a bookkeeper in Paris and moved to Barbizon to share a room with Jean and help Mme. Trotsky with the cooking and cleaning. Like van Heijenoort, she found the choice easy to make: she was a comrade, committed to the cause, committed to Van, and proud to be of service to the charismatic Trotsky.

It was a cozy situation. Because the Villa Ker Monique was much smaller than the houses of Saint-Palais and Prinkipo, the "live-in" core of comrades was kept to a maximum of six. Besides Lev Davidovich and Natalia, Van and Gaby, there was Sara Jacobs, an American secretary-typist who had joined the group in Prinkipo in the last months, and Rudolf Klement, a German secretary-bodyguard. Also, in contrast to the steady stream of visitors who had come to the villa at Saint-Palais, only Liova, his companion Jeanne Martin, and Henri Molinier visited Barbizon. Otherwise, almost no one, even those in the Paris group whom Trotsky was now seeing weekly, knew where he was living.

Again, Trotsky's whereabouts were kept secret for security reasons. More than anyone or anything, his supporters feared Stalin and his agents, and they were on a constant alert in that regard. At the same time, it was also official French policy to keep a low profile on Trotsky's presence in France. On both counts, there was anxiety by the cadre that Trotsky might be recognized and "found out." The distinctive look that had thrilled van Heijenoort in Prinkipo had now become a liability, and Van worried that the first passerby would identify Trotsky "because of that powerful gaze of his." Against this possibility, when the Old Man began taking regular trips from Barbizon to Paris on the public bus with either Van or Rudolf Klement accompanying him, Trotsky did take some pains to disguise himself, by shaving his goatee and by covering the lower part of his face. And, in spite of Comrade Van's apprehension, he did pass unnoticed among the other commuters.

Aside from the constraint of maintaining his anonymity, Lev Davidovich had not felt so free in years. He was at liberty to go wherever he wished. In Paris he met regularly with local leftist leaders as well as leaders of independent Socialist and Labor party groups who came from other European countries to discuss the goals and strategies of the Fourth International. After these meetings were over, he allowed himself the pleasure of a stroll down Boulevard Saint-Michel or some other street in the Latin quarter. With his son on one side of him and Comrade Van on the other, he would stop and look in the windows of the very bookstores where Jean as a student had gotten his first taste of Paris.

This, then, was the period of Comrade Van's greatest optimism, the time he had most in mind when he later said, "I didn't think Trotsky was a loser. I really thought we could do something." It was also the example he cited most often in stressing his point that one must never think of the thirties as a block but rather as a time of rapid change (sometimes from day to day and certainly from month to month), when for a short while he and his comrades thought propitious circumstances would aid their cause. In France Daladier was letting them be, and in Russia Stalin had not yet completely and firmly embarked on his ruthless crusade against every conceivable rival.

In this momentary quasi-favorable climate, the group began to explore wider possibilities for concrete action that would go beyond propaganda. As part of this new thrust, Van and Gaby left Barbizon for Paris in order to work more directly with the cadre there. Once or twice a week he returned to Barbizon to work with Trotsky on correspondence and current translations, and of course he saw his leader in Paris at the weekly meetings. Van Heijenoort later suggested that at this point there had been some overtures from comrades in the Soviet Union, but these were tentative rather than concrete, and nothing ever went beyond the idea stage.

Then, in 1934, a political storm raged through France that changed the mood of the country, and the balance of power shifted sharply to the right. In January of that year, Serge Alexandre Stavisky, a swindler associated with the municipal pawnshop of Bayonne, either committed suicide or was murdered by the police. Because Stavisky had connections with highly placed government officials, the financial scandal that ensued rocked France. The right accused the Radical Socialists, the party in power, of deep corruption; there were allegations that the latter had ordered Stavisky's murder to keep him from revealing names and details. In February, extremists staged bloody riots in Paris that Daladier had to repel with force. By the spring of 1934 Daladier was forced to resign, and the rightist Gaston Doumergue came in to form a new government. Many of the accused politicians were later cleared, but the damage to the Radical Socialist party was beyond repair.

A minor part of Doumergue's political inheritance was Leon Trotsky—a hand-me-down he would have liked to refuse. The new administrators would never have permitted Lev Davidovich to enter France, but for the time being they looked the other way, perhaps hoping that Trotsky would somehow disappear. Doumergue's ministers were unfamiliar with Trotsky's dossier and ignorant of the details of the arrangements and stipulations that had been made by Daladier. What happened as a result of all this non-information was that in April 1934, with life imitating a classical French farce (albeit minus lovers leaping out of beds), Trotsky was uncovered in Barbizon.

Until that moment, Lev Davidovich had lived in the Villa Ker Monique for six months, during which time the mayor, the police, and all local officials remained completely unaware of his presence. He and his wife had gone for occasional walks in the nearby woods and, on rarer occasions, had strolled through Barbizon with Van, but the Trotskys had had no contact with the locals. Van and Gaby, the only French members of the group, did the shopping, went to the laundry and the post office, and tried to present themselves as run-of-the-mill young newlyweds.

Barbizon, however, was a very small town, and the household at the villa, composed of six adults some of whom looked foreign and none of whom went to a daily job, could not fail to attract attention. One detail in particular piqued the curiosity of the locals: since there were no visible children, why was the tall blond man buying such large quantities of milk? In the absence of data about the mysterious strangers, invention flourished. Much after the fact, van Heijenoort learned that the rumor with the greatest currency (and the one which amused him most) was that they were a gang of counterfeiters. The supporting evidence for this conclusion was precisely the inordinate amount of milk they drank, because, so the folklore went, milk is a beverage counterfeiters must drink in order to neutralize the toxic effects of the lead used in the printing of money. Comrade Molokan had been metamorphosed into Comrade Forger.

It is not clear why the usually obsessively careful comrades did not anticipate that their presence in the small town would inevitably raise eyebrows, lead to questions, and cause trouble. Soon enough, the police had their suspicions and began to watch the house. One evening, on a pretext—his headlight was out—they stopped Rudolf Klement, the German member of the Barbizon group, who was returning to the villa on a motorbike. Klement, with his heavy German accent, was not forthcoming about where he had been or what he was doing; moreover, the bike was registered in van Heijenoort's name, not his. The police used this irregularity as grounds to charge him with theft. He was arrested, searched, and the briefcase he carried was carefully examined and taken apart in the expectation that it would be lined with counterfeit bills. Instead of fake money, the police found literary currency: newspapers, journals, and a stack of letters from all over the world addressed to Comrade Leon Trotsky. Klement, as it happened, had been on his regular run to Paris to collect the mail and deliver it to the Old Man, but the first readers of this batch would be the *gendarmes.*

The flabbergasted district attorney of Melun (the *département* in which Barbizon is situated) called Paris to learn the conditions of Trotsky's residence in France and to get instructions about what action to take. He was told that Trotsky had a perfectly regular visa but that he was supposed to be in Corsica. And why Corsica? At first, it seemed to be pure whimsy on the part of the person answering the telephone. Later, however, van Heijenoort speculated that the notion may have originated with the press reports from Marseilles. When Trotsky's initial entry to France was under consideration, there were officials who had urged that his French visa be limited to Corsica. Having missed him because of Liova's subterfuge, perhaps one journalist, for lack of any better information, had guessed that he had indeed gone to that island.

In any case, after eight months of silence, the press was delighted to rediscover Trotsky—he always made good copy—and the story was sensationalized. Journalists seized upon the Corsican angle; there were editorials demanding that he go back there, where, of course, he had never been. The rightist newspapers wondered why Trotsky should be allowed to live in *any* part of France, while those in sympathy claimed his status was legal. The clamor, pro and con, grew heated and threatening. Although the riots in response to the Stavisky scandal had abated, much of the population was still in an ugly mood and ready for new protest and confrontation.

Everyone directly concerned, including Trotsky, his followers, the mayor, the district attorney, and the chief of police, considered it urgent that the Old Man leave Barbizon. Almost immediately, he and Natalia did so under the cover of darkness; no one saw them go. A few months earlier, Liova had had the foresight to rent a small house in Lagny, not far from Paris, and now they went there for safety.

Comrade Van was enlisted to remain in the villa and to behave as if the Trotskys were still there. Each morning he opened the windows and shutters of their bedroom, and in the evenings he closed them. A group of reporters hung around the gates of the villa; many people came simply to stare, and a few to demonstrate. Comrade Van was nervous, a bit amused, and more than a bit frightened. Comrade Trotsky had not given him a name to fit *this* assignment.

"To my astonishment the comedy succeeded," he later wrote in *With Trotsky in Exile.* "Nothing shook their certitude that he [Trotsky] was still in the Ker Monique villa. Shortly I began to read in the newspapers details that I had given by telephone the day before to Liova or Raymond Molinier, and I realized that the reporters had connected a field telephone to our line. It then became quite easy to deceive them. I would simply give fictitious details in a confidential manner over the telephone. As I had taken the name Marcel, the newspapers were soon full of Marcel's doings."

*The hostile crowd in front of the villa. Barbizon. 1934.*

One Sunday afternoon, several hundred people came by bus to stage a protest. They were all around the villa and it seemed to Van as if the woods behind the house had been invaded. There were shouted insults and threats. Two local *gendarmes* were stationed outside, but Van judged them helpless against so many. He was in the house with Benno and Stella, the two German shepherd watchdogs the group had acquired in Saint-Palais. (Stella and Benno were great favorites as pets as well as useful guards. Trotsky always had animals around him and took personal responsibility for their care.)

Noticing a man climbing the fence, Van, with Benno at his side, dashed outside toward him. When the invader shouted that he was in his own country and could do as he pleased, Van replied, "I am also in my country." The man on the fence had expected to find a foreigner and was taken aback by Van's perfect French. At the top, he stopped, disconcerted. The sight of Benno snapping and growling at him inspired prudence and he climbed back down on the other side. Nevertheless, it was

*Comrade Van joking with reporters in Barbizon. 1934.*

clear to Van that what had begun as a farce was no longer quite so funny.

While all this was going on, Trotsky moved a second time, from Lagny to Chamonix; Natalia went to Paris; and Henri Molinier, the great negotiator, was talking to Doumergue's people about where they would permit Trotsky to live now. It was at that point that "Marcel" told the reporters that his chief was long gone. He was relieved to see that they were not unduly angry with him for having tricked them; everyone seemed to agree it had been a "good fight." Then he, too, left the Villa Ker Monique and Barbizon and joined Gaby and the rest of the Trotskyite group in Paris.

With a certain embarrassment, van Heijenoort later confessed that he had rather enjoyed the game of hide and seek, but he also said, "During all the years I spent with Trotsky, this is the only time that I knew fear. Passions were high. I was staying there every night, all alone."

Whatever Van may have become accustomed to in the service of his cause, one thing he was *not* used to was solitude, and although from the moment he joined Trotsky he was prepared to give his life, he obviously had not envisioned that the gift might go to the ugly and angry crowd of Parisians who stood outside the villa crying for the absent Trotsky's blood.

* * *

The Molinier brothers searched and negotiated for two months, and in that period Trotsky moved five times. Finally, through a friend in the schoolteacher's union, Comrade Van found a house that both Trotskyites and government officials agreed upon as being secure and inconspicuous. True, it was far from Paris, in the village of Domène, near Grenoble, but it was not as far as Madagascar or Reunion Island, which were two of the places the government had proposed. The house belonged to Laurent Beau, a schoolteacher, one of the "politically sympathetic" people that someone in the group always seemed able to find and call upon for help. Beau was a Freemason, not a Trotskyite. With relief, Trotsky and Natalia settled into the Beau household and a routine was established. Comrade Van was to divide his time between the core group in Paris and Trotsky in Domène. There he would do his always excellent job of note-taking and letter writing as well as translating anything the Old Man happened to be writing.

Again, the loyal secretary was shuttling back and forth, this time more than three hundred miles each way, and as usual, he had no complaints. But his life had changed since Gaby had become his companion. Here and there he had a few stolen moments for his own purely personal pleasure. In all circumstances he was a man who, as he put it, believed in strong feelings, and he acted upon his feelings. In spite of his claim that he was timid, he was a full-blown romantic and never afraid to use the word love in all its ramifications. He loved Trotsky, but he was also in love with Gabrielle Brausch.

In October of 1934, Jean and Gaby did something which for them was surprisingly unconventional: they married. In their circle of radical comrades of the Fourth International, people had relationships like marriage, but to feel the need to have such a union sanctioned by church and state went against the grain. It was considered bourgeois. However, Gaby was pregnant, and that made a difference. "She wanted it because of the child, and I was not opposed," was van Heijenoort's later explanation of their departure from the bohemian ways of his group.

The following January, Gaby gave birth to their son, also named Jean van Heijenoort. As luck would have it, just at the time his son was born and for a week afterwards, Jean *père* was in Domène with Trotsky. Marriage and fatherhood were not supposed to and, indeed, did not in any way alter Comrade Van's dedication to Trotsky and the revolutionary cause. That mission remained his primary commitment and had first call upon him, mind, body, and perhaps soul. Moreover, Gabrielle too had become deeply involved in the work of the Trotskyite group in Paris. She again worked at a bookkeeping job in the daytime and devoted all her spare time to political activities. Thus, very early in his life, baby Jeannot was taken to Creil to be cared for and loved by his grandmother Hélène (by this time returned from Saint-Quentin), his great-grandmother Adèle, and the same aunt and cousins who had been responsible for his father's early upbringing.

Politics was at the center of Gaby and Van's life. Dedicated to global change, they were indifferent, if not downright hostile in their attitude towards domestic conventions and customs—notwithstanding their marriage for the sake of their son. It didn't matter to Gaby that she was the breadwinner while Jean brought no money into the household; he was doing important work for the movement she believed in. For the same reason, she did not complain about his nights out with the comrades. And it didn't matter that without advance warning her husband might take off to an unknown place for an unspecified amount of time; in his line of work, that was to be expected. They were both willing to do whatever seemed necessary, in small or giant steps, to forward their cause. There was no disagreement between them about that. Instead, they argued about what was most important to them: the details of their politics.

It was their way of life. Then, and ever after, the Trotskyites were known for their constant, compulsive, hairsplitting, ideological arguments about the precisely correct way to conduct the class struggle. In one of the more serious of these splits, in this case a disagreement between Raymond Molinier and Trotsky on a matter of strategy, Gaby took Molinier's position, as did Van, perhaps out of loyalty to his first mentor. But soon he reconsidered the issue and returned to Trotsky's camp, while Gaby persisted in supporting Molinier.

On a different issue, Gaby allied herself with Pierre Naville, the intellectual purist of the group, who was always in conflict with Molinier. After the Fourth International was formed, it was Trotsky's idea that it should be aligned with the Socialist party of France where there would be greater potential for influence and action. All their attempts to alter the official Communist party line had ended in absolute failure but perhaps the Socialists would be more receptive. Most of the Paris comrades agreed to follow Trotsky's recommendation, including, in this instance, Raymond Molinier and his brother Henri, but, as usual, there were dissenters, this time led by Pierre Naville and joined by Gabrielle Brausch, who was, once more, in sympathy with those opposed to Trotsky. Naville's group argued strenuously against the tactic and refused to abandon the notion that the members of the Fourth International should be pure Communists and nothing else: not Stalinists, not Socialists, but true Marxist Communists. This was a fracture, and there would be others, that not only loosened the fibers of the tightly knit cell but, inevitably, colored Gaby and Jean's personal relations.

It is difficult to resist the thought that when Comrade Gaby spoke with the voice of those who differed with Trotsky, at least some part of what moved her was personal. For all her staunch belief and willingness to work at any task for the good of the cause, did she not feel resentful at always playing second fiddle to Trotsky's demands? Did she not, for example, resent her husband's absence at the birth of their child and the week following when he remained with Trotsky writing letters? Gaby was an assertive, articulate person; if she felt some hostility towards the Old Man who always came first in her husband's heart, she had found one way to express it. A few years later she would do so more directly in a fight with Natalia Ivanovna.

During the year (1934–1935) that Trotsky lived in Domène, the political climate, from his point of view, deteriorated and polarization between the left and right increased. Personally, he and Natalia felt confined and uncomfortable in the Beau household. Trips to Paris were out of the question and even locally, his movement was restricted. Although he still had many visitors, now that the authorities both local and national knew *exactly* where he was, he and his visitors were under constant surveillance. Yvan Craipeau hid in the trunk of Raymond Molinier's car when he came to visit so the police would not know. It was hardly an atmosphere conducive to easy relations, and the strain began to tell. Trotsky always argued with his enemies, but now he began to argue with his friends. He asked his landlord to remodel the part of the house in which he and Natalia lived and then got into a bitter dispute over the costs. Natalia, in her memoirs, would mention the Beaus as "excellent people," but Trotsky wrote (in his diary), "There is no creature more disgusting than a *petit bourgeois* engaged in primitive accumulation. I have never had the opportunity to observe this type as closely as I do now." Comrade Van also got his share of sarcastic invective. When Trotsky found a receipt for a drink Van bought on the train from Paris, which he had carelessly left in a book, the Old Man waved the bill at him and shouted, "Oh, oh, you went on a binge in the dining car!"

The Barbizon calamity had foreshadowed the change. Now there were further signs that Trotsky's situation in France was precarious, and no indications that favorable winds were coming his way. Something had to be done. While the Molinier brothers sought yet another safe house in France, others among the comrades explored the possibilities for escape and legal or illegal refuge in some other European country. Van Heijenoort went to Holland and Belgium to test the political waters there and came back with a favorable response from sympathizers in those countries. A plan was devised that included a scheme for obtaining a reliable false passport. Supporters in each of those countries were prepared to help at the crucial moment. In the end, however, none of the illegal maneuvers was necessary because Norway, which in 1935 had elected a Socialist government, came forth with an offer of haven. Oslo would be the next stop in the odyssey.

Chapter Six

# Herring, Norwegian and Red

France and Norway: 1935–1936

*Erwin Wolf and Jean van Heijenoort in Antwerp, after their arrest and imprisonment in Norway and Denmark. 1936.*

When Trotsky was out in the garden, he went around with his walking stick, knocking it against the fence. I asked him why he did it, and he said, "It is an old habit from my years in prison."

GUNVOR WRAAMAAN, *1989*

AFTER SOME UNSETTLING last-minute hesitation by the Norwegian government, the departure from France proceeded without a hitch. Relieved to have Trotsky off their hands, the French authorities did their best to facilitate his departure; even the police were "extremely discreet," van Heijenoort noted in his memoir. "The whole trip was a much simpler affair than Trotsky's previous journeys." On June 13, 1935, in the Gare du Nord, Trotsky, Natalia, Comrade Van, and one other comrade who came to help with guard duty boarded the train for Antwerp. There they were met by Jan Frankel who, by pre-arrangement, had come from Czechoslovakia to join them. Frankel replaced the French guard, and the exiles went on to Oslo by ship. When they arrived, the immigration proceedings were routine and no reporters were present. Trotsky and his party simply walked down the gangplank in a group with all the other passengers.

Considering the difficulties of the previous year, Trotsky's entry into Oslo was almost unbelievably easy. To a large extent this was due to the good work of Walter Held, Trotsky's "minister" in Norway who was himself a refugee from Germany. He and his Norwegian wife were part of the small but effective network of European Trotskyites and "fellow travelers" who, like guardian spirits, appeared at crucial moments to offer a lifeline. The Helds and various friends had made all the necessary arrangements for visas and, because their contacts were exceptionally good, they had found a remarkable place for Trotsky and Natalia to live.

Immediately upon arrival the exiles were taken to Hønefoss, a small town forty miles from Oslo, where they were to live as guests in the house of Konrad Knudsen, a newspaper editor and later a Socialist member of parliament. This dignified and protected situation was a far cry from their recent clandestine arrangements in France. Nevertheless, there was a problem: the Knudsen house was not large and there was not enough room for anything approaching the usual cadre of secretaries and bodyguards. As a result, Trotsky decided that only one person would remain with him in that capacity. The job fell to Jan Frankel and, quite unexpectedly, van Heijenoort returned to Paris to continue his work with the Trotskyite group there.

For the first time in three years, Van was no longer in close personal contact with Trotsky and reporting to him on a daily or twice-weekly basis. The twelve hundred miles and several countries that separated him from the Old Man's "powerful gaze" had a liberating effect. Freed from Trotsky's constant beck and call, he felt ready to assert himself. An opportunity to do so presented itself almost immediately, because when he returned from Norway, he found his comrades in the midst of a factional fight.

When, in 1933, Trotsky and the former members of the *Ligue Communiste* decided to call themselves the Fourth International and, later, to join the Socialist party in the hope of having broader influence, a splinter group headed by Pierre Naville had resisted. Eventually, Naville's group also joined the Socialists, and although they continued their infighting on small points, the two Trotskyite factions found themselves working together on larger issues. In the fall of 1935 (after Comrade Van returned from Norway) the Socialists, worried about the growing power of the Trotskyites, expelled both groups. Closing ranks, the two Trotskyite factions re-formed as one, and again tried to increase their number and enlarge their sphere of influence. But in Raymond Molinier's opinion, little progress was being made in this direction. One of the main reasons, Molinier argued, was that the writing in *La Verité,* the official journal of the Fourth International, was far too abstract and way over the heads of non-intellectual readers. How could they hope to attract "the workers" with something like that? Van Heijenoort, Gaby, Jeanne Martin (formerly Raymond Molinier's wife, now Liova Sedov's companion), and a few others agreed and urged a change of attitude and style. When after a time it became evident to them that their pleas and exhortations were having no effect and that *La Verité* was not changing, they broke away from Trotsky, Liova, and Naville and started a journal of their own titled *La Commune,* written in more popular language and designed to appeal to the ordinary worker.

Once again, the disagreement was mostly about tactics, not ideology, and therefore not terribly important—except that in a group so small, they could ill afford to divide their strength and any fracture was serious. Furthermore, in this instance there was the additional tension created by the Molinier–Martin–Sedov triangle. Three years before Jan Frankel and Van's arrival in Prinkipo, Raymond Molinier and Jeanne Martin, then man and wife, had been part of Trotsky's staff at the time when Trotsky's son Liova was also in Turkey. Trotsky found Molinier to be a "practical, energetic and obliging young man." Even in Turkey Molinier had been successful in finding a house for the Trotskys. While Trotsky praised Molinier as "the prefiguration of the future communist revolutionary," Liova fell in love with Raymond's wife. Threatening to kill himself if she would not live with him, Liova persuaded Jeanne to leave Molinier.

Although willing to change partners, Jeanne insisted upon her political independence. Thus, on the *La Commune* issue, she agreed with her ex-husband and joined those opposed to Liova and Trotsky. For different reasons, both Liova and Trotsky were furious: Liova, because he felt betrayed personally as well as politically, and Trotsky, because he had strongly disapproved of the liaison between his son and Jeanne Martin. He had counseled and argued vehemently against their living together because he was sure it would lead to trouble. Now, it gave the Old Man small comfort to have his prediction come true.

But if Trotsky was able to say, "I told you so" to Liova in regard to Jeanne Martin, he had had no intimation that his ever-faithful Comrade Van would wander from his side and line up with the opposition. Based upon the letters Trotsky received from all the parties involved in the controversy—including Van—the Old Man came to the conclusion that his youngest disciple was not only defecting but also trying to deceive him. The fact that Trotsky was wrong on the second count was all the more wrenching to Van. He was torn between his filial and quasi-religious devotion to Trotsky and his agreement with and admiration for Raymond Molinier, his first mentor. Here was the first portent of a situation that would repeat and cause him no less anguish with each repetition.

When it turned out, after a few months, that *La Commune* had no more success than *La Verité* in luring the working class into the arms of the Fourth International, van Heijenoort abandoned what now seemed to him "an aimless adventure." There followed a short interim when he belonged to neither of the two factions of the group. "I drifted," he later said, speaking directly of politics and, less clearly, about his emotional detachment. But in a practical sense, he did quite the opposite. On his own, in the *real* world, almost for the first time in his life, without a scholarship from a *lycée* or bed and board from Trotsky, he decided to look for work. Gaby, as always, had her job as a bookkeeper; but since he was no longer occupied as Trotsky's secretary and was not directly involved in the Paris cell, there was no reason she should support them both.

Van Heijenoort applied for a position with La France Mutualiste, a large insurance company, where a Socialist acquaintance, André Thirion, held a high-level administrative position. Van made a powerful impression upon his boss. In his book, *Révolutionnaires sans Révolution,* written at least thirty-five years later, Thirion gives a vivid picture of the twenty-three year old van Heijenoort, applying for the job:

> A young Nordic type, very tall, a perfect Aryan according to the Nazi canon, fresh from his experience as a student in mathematics, came to ask for work. His name was van Heijenoort; he was one of Leon Trotsky's secretaries. He had entered into the service of the great man as one enters religion. Modest, gentle but uncompromising and persevering, intelligent, hard-working, living on very little, he was sort of an angel-slave whose "revelation" preserved him from impurity, hesitation and doubt, and filled him with ineffable joy. He brought two or three other Trotskyists with him, Craipeau and a frightful, twenty-year-old harpie who was hopelessly in love with him.

Asked many years later about the "harpie," van Heijenoort said he could not remember who she was, but the rest of the passage, he modestly conceded, was accurate.

La France Mutualiste needed an actuary, a person who knew integral calculus. Van Heijenoort, with his years of mathematical training, knew calculus and much more. He was hired immediately and, of course, did an excellent job, because he was simply incapable of being less than meticulous in any kind of mathematical or abstract intellectual work. All the same, his devotion to actuarial problems did not distract him from his mission as revolutionary any more than marriage or fatherhood had done.

Soon after Comrade Van began working at the insurance company, he wrote to Trotsky to tell him that he had left Molinier's group and had revised his opinion regarding the idea of *La Commune*; he now saw that its orientation was "foreign to Marxism." He added that he had rejoined the original circle. Trotsky, with great warmth, answered immediately: "When the prodigal son returns, a lamb is cooked in the paternal home."

Van Heijenoort's return to the fold, based upon what he described as a revision of his temporarily dissident position, may have been an honest intellectual choice, but it was also enormously important to him emotionally. The father–son rupture was mended and he was in Trotsky's good graces again. But while stretching to repair one tear, he created another. His wife, Gaby, did not share his revised views about *La Commune*; when Van returned to his "father" and the original group, she remained in the *Commune* camp with Raymond Molinier and Jeanne Martin. Although the Gaby–Van–Trotsky triangle was somewhat different from the situation between Jeanne, Liova, and Raymond, the passions and tensions surrounding their entanglements were very much the same.

From any angle one chooses to look, the filial–paternal relationship between Jean van Heijenoort and Lev Davidovitch dominates the picture; the filial half of the equation is mentioned first because it was Van who had most at stake in maintaining the relationship and who, for most of their years together, was willing to give up a great deal of independence to perpetuate it. Where Trotsky was involved, there was no middle ground, no compromise. One was with him or against him, and since the young man's need was stronger, and qualitatively different, it seems he was always compelled, for one reason or another, to select Trotsky's way. On the occasion of the *Commune* rift, his return was clearly a relief to both of them. Although Trotsky's need for his young comrade secretary was much less a father–son affair and much more a very practical need for the quality of Van's intelligence and service, that difference was a minor issue, if indeed it was consciously registered by either of them at that point.

In the inner circle of the Fourth International not much of what was going on emotionally was considered worthy of examination, private or public. Here however, it does deserve some consideration. During Leon Trotsky's eleven years of exile, none of his *aides-de-camp* served him longer or more loyally than Jean van Heijenoort. Within the closest circle, only Trotsky's devoted wife, Natalia Sedova, spent more nights under the same roof. Leon (Liova) Sedov, their son, who had come with them when they were first sent into exile in Turkey, left after two years to go to Berlin; Jeanne Martin went with him. By 1933 they had returned to France but Liova never again shared a house with his parents.

Although Liova was usually second in command in party matters and in charge of everything having to do with Trotsky's personal arrangements, relations between father and son were volatile and fraught with difficulties. Jan Frankel, the other secretary closest to Trotsky, had spent almost three years in Prinkipo before van Heijenoort arrived to replace him. By the time the exiles landed in Mexico, in 1937, Frankel's tolerance for Trotsky's temper would be exhausted, and he, too, would separate himself from the Old Man, physically, and eventually politically as well.

After the initial excitement and glamour subsided, living and working with Lev Davidovich could be a hardship assignment for anyone. The ever-present threat of assassination or abduction meant constant preoccupation with security in and out of whatever house the group happened to be occupying. Only rarely were there enough people to do a proper job of guarding, a task that even under ideal conditions would be bound to create tension. Their rule was that Trotsky was never supposed to go anywhere, unless he was accompanied by at least one gun-toting bodyguard. Moving from one place to another, whether it was a short trip or a major change of residence or exile in a new country, always required a carefully engineered set of plans and alternatives. Still, these were temporal, logistic problems which had concrete solutions.

Other, interpersonal, problems that lay beneath the political ideology of the group were much more difficult to solve, partly because they were never overtly acknowledged. Here was a circle of intelligent, dedicated, opinionated men and women, all ardent believers in their cause, all comrades, all of supposedly equal status, all working full time without pay because changing the world would be reward enough. Apostles, in other words. Their leader was a great man, a Russian, with an ironic turn of mind and a biting wit, who did not suffer fools. Together they were engaged in an epic struggle—all uphill. When things went wrong, as they often did, and tensions arose, where else would the comrades vent their spleen but "at home" and upon each other?

The outside world looked upon the Trotskyites with scorn and outright hatred. They were reviled on the left by the Stalinist-Communists, on the right by the conservatives; in the center, the liberal socialists didn't much care for them either. Once more Van had cause to describe himself as a pariah. "We were political outcasts; *comme des chiens lepreux* [like leprous dogs]." Having so many enemies naturally drew the Trotskyites together, but at the same time, among themselves, they had to make every small difference of opinion perfectly clear. Like members of a high-strung family, they had a positive compulsion to disagree. Their passions ran high in the name of "isms" but bubbling and boiling beneath the surface were the ordinary human passions, the love triangles and the personality conflicts that added fuel to the fire.

Van Heijenoort, in his early twenties, already had long experience in containing emotion. In an atmosphere that would have driven most people to distraction, he thrived. He later claimed this period as "one of the happiest times in my life." The alien had found his place in the world, however impossible and risky it was. He was doing something to complete himself. An emptiness was filled and a father resurrected.

* * *

In the spring of 1936, following the election of the Socialist Léon Blum, a wave of strike activity swept across France. Demanding wage increases, a forty-hour work week, and holidays with pay, the workers made use of a new tactic: the sit-down strike. Instead of using the traditional form of demonstrations *outside* factories and offices, the strikers remained *inside* the buildings but refused to work. The employees of the France Mutualiste—van Heijenoort, the actuary, among them—voted to join the general strike and they, too, occupied their building on a twenty-four hour basis. Van Heijenoort was elected head of the France Mutualiste strike committee, which chose to do its sitting-in in the director's office. Meanwhile, André Thirion, Van's Socialist boss, took advantage of the fact that he was unable to perform his regular duties and spent his time organizing the workers in a rag factory where, he writes, "Working conditions were abominable, like something out of a report on Shanghai or a Zola novel." By June the Blum government had brought the unions and management together and the strikes were settled largely to the satisfaction of the unions. At France Mutualiste, Comrade Van sat at the big conference table and negotiated the final settlement for his constituency.

In the midst of the sit-in, Trotsky sent Comrade Van his congratulations for the good work he was doing as the leader of the strike committee, and, in the same mail, a manuscript for Comrade Van to translate. The title was *The French Revolution Has Begun*, and Comrade Van, seated at the desk of the director of France Mutualiste, finished the translation before the strike was over.

And now, a brief period of unusual calm: relations between Comrade Van and Lev Davidovich were back on keel. Van Heijenoort resumed his job at the insurance company and was also active in the Trotskyite cell in Paris. There was even a temporary reconciliation between the split factions. In Norway, Trotsky and Natalia were safely ensconced in Konrad Knudsen's house in the seeming tranquility of the countryside.

Gunvor Wraamaan, a piano teacher who lived in another part of the Knudsens' house, used to sit in the garden with the Trotskys and talk about music. She thought they were kind, sensitive people; she knew of their trials and felt sorry for them. "When Trotsky was out in the garden," she later recounted, "he went around with his walking stick, knocking it against the fence. I asked him why he did it, and he said, 'It is an old habit from my years in prison.' " The Trotskys invited Gunvor and her family to his birthday party. She remembered the cake and the writing upon it: "Long live the Fourth International!"

As Trotsky bided his time, he continued to write. *The Revolution Betrayed* was finished in Norway, and, keeping abreast of what was happening in Russia, Germany, France, Spain, and the rest of Europe, he wrote political tracts and articles. At the outbreak of the Spanish Civil War, in July 1936, Trotsky considered a secret move to Catalonia—thinking that was a potential theater for action.

Meanwhile a storm of terror was gathering in Moscow. The shift from fair to foul came on much more abruptly in Norway than it had in France. There, a year had elapsed before the comrades decided it was essential to find a way to leave. But in Norway, after so felicitous a prologue, an accelerated and more dangerous version of the French scene was to be played out. In this case the motivating factor was not a change in the Norwegian government. Instead, it was external pressure and threats from the Soviet Union, where Stalin had unleashed the first of the infamous treason "show" trials directed against his former comrades. The accusation was made that the murder of Sergei Kirov, a member of the Politburo, was a Trotsky-inspired conspiracy to seize power. Although future historians would reveal that Stalin himself had ordered Kirov's assassination because he feared that Kirov was becoming too popular and too powerful, at the time even Stalin's enemies were unaware of the facts.

When the news of the Moscow trial reached Paris, Liova Sedov became very concerned about his father's safety. And when he learned of a peculiar theft at the Knudsens' house, his anxieties multiplied. On a day when the Trotskys and their hosts were away on an excursion, vandals had broken in and stolen some of Trotsky's letters and documents. The thieves were arrested and discovered to be a group of pro-Nazi Norwegians. A few weeks later, the thieves were brought to trial and Trotsky was called to testify as a witness against them.

Liova feared that Stalin's powerful arm was poised to reach into Norway, and at that moment, Trotsky had only Erwin Wolf (a Czech, who had replaced Jan Frankel) living with him as secretary–bodyguard. Because Liova did not have a French passport, his ability to come and go from country to country was restricted and complicated. In any case, Comrade Van was considered the most able when it came to matters of security. It was Van, therefore, who was immediately sent to Norway to provide whatever extra protection he could muster for the Old Man. In retrospect, Van said, neither he or Liova had had any foreknowledge of just how perilous the whole venture would be.

Traveling by ship from Antwerp, Comrade Van arrived in Norway on August 25, 1936, three days before the trial of the vandals who had stolen the papers. As his ship was docking in Oslo harbor, he read the newspaper headlines announcing the executions of Zinoviev, Kamenev and fourteen others who had allegedly confessed to participating in a Trotsky-inspired plot to overthrow Stalin. Now, there was no longer any doubt that Trotsky was in very grave danger. By the time of Trotsky's day in court, it was clear that the Norwegian government's shift in attitude was in response to pressure from the Soviet Union that something had to be done about Trotsky, or else. The "or else" was the threat that unless Trotsky was arrested and either jailed or deported, the Russians would cut off their large imports of herring—not a small matter to the Norwegians. The government responded quickly.

In the courtroom, to everyone's astonishment, the proceedings took a bizarre, one-hundred-and-eighty-degree turn. Trotsky the witness found himself in the role of Trotsky the defendant, and under arrest by the Norwegian government. The moment the trial ended Trotsky and Natalia were placed under house arrest at the Knudsens', and van Heijenoort and Erwin Wolf were picked up and taken to the police station in Oslo. There the bodyguards were told that unless they signed a statement saying they were leaving Norway of their own free will, they would be deported to Germany. They refused, and the next day, accompanied by two policemen, they were taken by train to Sweden. At the border, two Swedish policemen replaced the Norwegians and escorted the prisoners to Denmark where they were handed over to six Danish policemen. In Copenhagen, van Heijenoort and Wolf were put in jail, in separate cells, and all their clothes and all their personal belongings were taken from them.

Van Heijenoort later described the experience of having everything taken away as one of the worst things that ever happened to him, quite different from the fear of the crowd he had experienced in Barbizon: "You have no idea how it feels, to be completely naked in a cell, without even a blanket, to have *absolutely nothing* that is yours."

Material possessions were never important to Van, then or later in life; he did very well with little, but zero was unimaginable. He was a man who hated to be physically cold, but what he felt was something that went much deeper than being without a blanket or clothing. In the Danish prison, he felt stripped of every last shred of his identity.

To their enormous relief, the following day, Comrades Van and Wolf were put on a freighter—with no policemen—bound for Morocco, but not until they were underway did they learn they would be allowed to disembark in Antwerp. Until that moment, they had been kept in the dark as to where they were being taken and by whom. From Antwerp, the Belgian police, who were awaiting their arrival, escorted them to Paris, where finally they were released. Although van Heijenoort had not believed that the Norwegian government would actually send them to Hitler's Germany, he was mightily relieved to be home.

Meanwhile, Trotsky and Natalia were moved from the Knudsens' to another village, south of Oslo. For the next four months they lived on the second floor of a small house and, according to Natalia Sedova's report, were guarded by "twenty drunken policemen who sang obscene songs day and night." They were not allowed to receive visitors, with the exception of a Norwegian lawyer and Gérard Rosenthal, Trotsky's Paris lawyer. Through these two, Trotsky tried to institute proceedings in other European countries to refute the charges of treason that had been made against him at the Moscow trials. He was prevented from doing so by Trygve Lie, then the Norwegian minister of justice and later Secretary–General of the United Nations. In a move sponsored by Lie, the Norwegian government issued a special decree forbidding an interned alien from undertaking any court proceedings. He was forbidden to make public statements and his mail was restricted and censored.

Outraged, van Heijenoort would later write: "In its treatment of Trotsky, the 'socialist' government of Norway stooped to ignominies that, even in the dark days of Domène, the Doumergue government had never descended to." Irving Howe, in his book *Leon Trotsky*, also pointed to the tragic irony of Lie's actions by quoting Trotsky's own scornful and prophetic response to his imprisonment and the restrictions imposed upon him:

> This is your first act of surrender to Nazism in your own country. You will pay for this. You think yourself secure and free to deal with a political exile as you please. But the day is near when the Nazis will drive you from your country, all of you, ...

History records that Trotsky was correct. Four years later the Nazis invaded, and Norway fell.

With Trotsky muzzled in Norway, once van Heijenoort was safely back in Paris he spent most of his time and energy working with Liova on efforts to reveal what had really happened in Moscow. One of the results of their collaboration was the *Livre Rouge*, the first systematic exposure of the frame-up of the Zinoviev–Kamenev trial. Liova wrote it in Russian, while Van translated, proofread and saw to the printing. In a further attempt to clear Trotsky's name, they also worked with Gérard Rosenthal, Trotsky's lawyer, and a commission of inquiry that had been formed to investigate Stalin's charges. André Breton was a member of this commission, and thus Van had the opportunity to meet the surrealist whose works he had read and admired in his days at the *lycée*. He was very impressed by Breton as a political activist, and surprised that he always signed his name in green ink, adding "writer" underneath the signature. In the context of their meetings, it seemed inappropriate to Van to tell Breton about the Sunday mornings, five years earlier, when he had stood in front of the stove at the bookstore on Boulevard Saint-Michel, peering voyeur-like into the uncut pages of poetry. But he was to have another chance later, in Mexico.

Meanwhile, as the Trotskyites in Paris were doing their best to prove to the world that Stalin's accusations were false, Walter Held and others in Norway were desperately seeking a means by which Trotsky might put more distance between himself and his relentless enemy. Held, who in the first instance had paved the way for Trotsky's acceptance in Norway, was now helping in his flight. Never was the worldwide network of the Fourth International more important than at this moment, and never did the task of finding refuge appear to be quite so hopeless. But again, an "angel" appeared, this time in the guise of the great Mexican muralist Diego Rivera, who was a Trotsky admirer. The path was circuitous but the connections worked with the crucial assistance of Max Shachtman, a New York Trotskyite of many talents who had been to Russia and had visited Trotsky in Prinkipo, and Anita Brenner, a writer who had lived in Mexico in the 1920s and been part of the circle of artists and writers there.

Rivera had painted Trotsky's portrait at the Trotskyite headquarters in New York and had also put him into his controversial mural in Rockefeller Center, all without ever having met the man. In Mexico, Rivera gave lectures about art and its role in revolutionary culture and got into public arguments and debates with his rival, the painter David Siqueiros, a member of the pro-Stalin Mexican Communist party. When a cable came from his friend Anita Brenner, in New York, asserting that it was a matter of life or death to find a haven for Trotsky, Rivera jumped to attention. With his flair for the dramatic, he sought out the Mexican President Lazaro Cárdenas, driving over the bumpy and dusty backroads of northern Mexico where Cárdenas had gone to check on the land-reform program. On the spot, Rivera presented Cárdenas with the petition requesting that Leon Trotsky be offered asylum, and, on the spot, the President signed, with the condition that Trotsky promise that neither he nor his group engage in Mexican politics or internal affairs.

On December 19, 1936, Trotsky and Natalia, accompanied by only one guard (who went along not to protect them but to assure that they actually left the country), sailed for Mexico aboard the steamship *Ruth.* Like France, Norway had put no obstacle in the way of the exiles' departure. Far from it. Because of the constant threat from Soviet Union, the Norwegians were delighted to be rid of them. By the time the news reached Paris, the Trotskys were already at sea. Liova wasted no time; he asked—in fact, urged—Van to leave immediately and join them in Mexico. In nine days, he was gone, leaving Gaby and his little son Jeannot behind.

Chapter Seven

# A Blue House on a New Planet

Mexico: 1937

*Van, Trotsky, and the Russian secretary, at work in Coyoacán, 1937.*

We didn't know anything about what we would find in Mexico. There was great uncertainty. It turned out to be very good—with Frida, with Diego, with Cárdenas—but we didn't know.

JEAN VAN HEIJENOORT, *1983*

THE OCEAN LINER *Empress of Australia,* bound for New York, left Cherbourg harbor on December 28, 1936, with Jean van Heijenoort aboard. From New York he was to go to Mexico City, where he would join Trotsky and Natalia and help to reconstitute and organize the nucleus of European Trotskyites who had made their way to the New World. More than four years had elapsed since Van had abandoned his studies and seized the opportunity to devote himself to politics—four crucial years, during which he had lived in a state of alert, constantly wary, always watchful, employing all his resources and cunning. He had even learned to make an asset of his mild, innocent look and timid ways. With a shy smile he had falsified passports, used aliases, and tricked reporters into believing Trotsky was where he was not. On one mission or another, he had traveled from Paris to Turkey and back; to the various corners of France; to Germany, Norway, Holland, and Belgium. He had been in jail more than a few times. In short, Comrade Van was now a man of the world, and he saw himself that way.

The biggest change in van Heijenoort's personal life was that he had a wife and a son, but being a family man did not affect his decision to leave France. He simply told Gaby that Liova had asked him to go to Mexico. He and Gaby did not discuss the pros and cons of the assignment or whether or not he ought to accept. Such discussion was unnecessary; it was understood between them that "the cause," the mission, and his loyalty to Trotsky always came first. For Comrade Van, to shirk this duty for personal reasons would be, like flight in battle—unthinkable. In later years, van Heijenoort acknowledged the pain of leaving his family and the uncertainty he felt about the future, but he also said that if he had not been asked to go to Mexico he would have been sorely disappointed.

Aboard the ship to America, the "innocent," fair-haired van Heijenoort was befriended by an older man who was very interested in his political ideas. Almost immediately, Comrade Van concluded that the man was a Communist and that he was being sized up as a potential recruit for the Communist party. Taking up the game, the young Frenchman played his part well, keeping quiet about his connection with Trotsky, while making knowledgeable and generally sympathetic remarks about Marxism. The dialogue continued in this fashion throughout the journey, with the Stalinist acting as mentor and telling Van all the things he thought he ought to know. As the ship entered New York Harbor, Van learned that he would need United States currency in order to pass through immigration without problems. Shyly, he asked his "friend" if he might borrow a hundred dollars just to show to the immigration officers, and the man agreed. As soon as the formalities were over and van Heijenoort was safely on shore, he returned the money with many thanks and some vague words about keeping in touch.

At the time, this maneuver gave the young van Heijenoort a great deal of satisfaction, for he was certain that the Stalinist had had no inkling of his affiliation. Years later, in the wake of the Alger Hiss trial, Van learned much more about his benefactor. The fellow-passenger turned out to be Paul Massing, a high-ranking agent in the GPU, Stalin's secret police. Reveling in the irony of their shipboard encounter, he said, "Just think what would have happened to that guy if his GPU superiors had found out that he helped Trotsky's secretary get into New York."

So it was indeed a changed, *cagy* van Heijenoort who sailed into New York, and, as he would point out ever after,

> It was a changed world. That understanding is crucial. 1932 was pre-Hitler, and there were still hopes in Germany and in Russia. I was optimistic in 1932, even with Stalin in power, because things had not yet completely settled. But by the end of '36, when I was going to Mexico, we had already had the Moscow trials; we had Hitler; we had the Spanish civil war. It was a somber period. The situation was different, and the feeling was different. I was much less optimistic.

Another factor contributed to van Heijenoort's lost optimism. He had become a leader with ever increasing responsibilities. In Turkey, he had not been "part of the outfit," but by the time he arrived in Mexico, he was integrated into the tight little group of six or eight who were making the decisions. As a leader, he had many more specific things to worry about. This pessimism was to stick, in spite of the fact that from the moment of their arrival in Mexico, the Trotskyites were treated with the greatest warmth and kindness they had ever encountered in their many years of exile.

In New York, van Heijenoort spent a few days with the leaders of the American Trotskyite party exchanging information. From them he learned where Trotsky and the group were to live in Mexico. Then, in a snowstorm at the beginning of the new year, 1937, he boarded a flight scheduled to go from Newark to Mexico City. It was his first plane trip and he later recalled it in vivid detail:

> It was snowing, and I didn't think we would take off, but we did. It was a small plane, but there was a real bed, with sheets. When it got dark, I actually went to sleep, in my bed. I had even put on my pyjamas! Then I woke up with the feeling that the plane was not moving. We had landed in Memphis, grounded because of the blizzard. So everyone got off and I took the train. I remember riding for two days, looking at the ice on the telegraph wires along the tracks until we got to Brownsville, Texas, and there, the sky was blue.

In Brownsville, Van had time to buy himself a new suit. It was an outfit he would wear for years, whenever he wanted to look "dressed," and from the evidence of photographs, he did indeed cut a dapper figure. In his new clothes, Van boarded another airplane, flew to Mexico City, and took a taxi from the airport to the suburb of Coyoacán.

Natalia and Lev Davidovich's itinerary was completely different from Van's. They left Norway December 19 and traveled on the proverbial slow boat directly to Mexico. Van Heijenoort

*Jean van Heijenoort newly arrived in Coyoacán, wearing the suit he bought in Brownsville, Texas. 1937.*

left France nine days later and traveled via New York by ship, airplane and train. To the amazement of all, he arrived at the house on Avenida Londres, where they were all going to live, exactly one hour after the Trotskys.

The famous blue house on Avenida Londres in Coyoacán, a suburb of Mexico City, is presently the Frida Kahlo Museum and contains a collection of her paintings as well as memorabilia of her life. Her father, Guillermo Kahlo, a photographer, built the house and Frida was born in it. When she married Diego Rivera, he bought the property from her father, and the two artists lived there for a short time. They filled the house with paintings and a great collection of Mexican and Indian folk art, but by the time of Trotsky's arrival they had moved to their new, starkly modern, "linked" houses in nearby San Angel. So the charming Coyoacán house was available, and the generous Rivera offered it to Trotsky and his entourage saying he would be honored if the great man and his followers would live there for as long as they desired. The offer was eagerly and gratefully accepted.

"A new planet," was Natalia Sedova's summing up of the glories that befell them in Mexico. Although Comrade Van agreed with the poetic spirit of the phrase, his view was more terrestrial: "Mexico for me was what Abyssinia was for Rimbaud, a country that was outside the civilized, industrialized Western world." As in Turkey, van Heijenoort interpreted the clear skies, the warm Mexican sun, the lush color of the exotic plants as signs of the good that was to follow. And he was right, for, in contrast to Turkey, this world seemed to be populated by exuberant and sympathetic people.

While Van had traveled through a blizzard to come to this new "Abyssinian" land, the story of the Trotskys' arrival, like the Barbizon episode, was pure farce. After three weeks at sea with their one Norwegian policeman, the couple had finally sailed into Tampico Harbor where they were met by various government officials and invited to come ashore. Still affected by the trauma of their recent imprisonment in Norway, they refused, and said they would not disembark "unless they saw the faces of friends." Just as the frustrated Mexican authorities were about to back up their invitation by literally picking them up and dumping them on shore, another government boat approached with a welcoming party on board which included some of the familiar faces the Trotskys had been expecting: Max Shachtman and George Novack, the American Trotskyites, as well as several local and federal authorities and journalists. Frida Kahlo had also come, representing her husband, who was in the hospital with serious kidney problems and under doctor's orders not to leave his bed. Rivera was terribly disappointed to miss the actual moment of Trotsky's arrival, since he had done so much to make it possible. As Trotsky was to write later, "It was to him above all that we were indebted for our liberation from captivity in Norway."

The red carpet was rolled out, but carefully and not quite all the way, because the wary Trotskyites were concerned about what the Mexican Stalinists might do. President Cárdenas had sent a special train to Tampico to carry Trotsky and Natalia and the welcoming party to the capital, but because of Trotsky and Natalia's worries about too much visibility, the train left Tampico at night and arrived at a small station on the outskirts of Mexico City in the early morning darkness. Meanwhile, a phony welcoming party was being staged at the main station, and another celebration, with none of the principals present, was taking place at Rivera's house, while Rivera himself was still in the hospital. At the last minute, the doctors granted the great painter a temporary reprieve from his hospital bed so that he could take his rightful place at the genuine reception. When the train pulled into the suburban station, there was Rivera, after all, waiting to embrace Trotsky; and, in spite of all the subterfuge and the smoke-screen parties, so was a small crowd of photographers and reporters, ready to record the big event. No one knows whether the scheme worked to confuse the feared Stalinist agents, but if they too were present at the small station, they did no immediate or visible harm.

Never during his years in exile, and probably seldom at any other time in his life, had Trotsky experienced the effusive hospitality he was accorded in Mexico. Although no one in the leftist Cárdenas government supported him officially, there were many politically powerful individuals who were sympathetic to Marxist ideology and, in particular, to Trotsky's point of view. On the other hand, there were also many Stalinists in high places; as always, this was of the greatest concern to the exiles, and now most particularly so, as the Moscow trials were continuing. Nevertheless, the Mexican political environment was more protective than the newly arrived refugees had imagined possible. On a more personal level, Trotsky responded to Diego Rivera's expansive generosity with unguarded warmth and an unusual lack of restraint. His response to Frida Kahlo's charms was also immediate and powerful and would soon become a matter of concern among the comrades.

* * *

*Frida Kahlo and Diego Rivera. 1937.*

There is no scene more characteristic of the relationship between Comrade Van and the Old Man than the hour of their arrival in Mexico. They had not seen each other for five months, not since their separate arrests in Norway when they had both been in very serious danger; and they had both been on a long, if not arduous journey. They therefore indulged in a brief show of affection, after which Van reported the news from Paris. Half an hour later, Trotsky was dictating a letter to Van, and others were looking around for a Russian typist and a typewriter. If van Heijenoort wanted a greater demonstration of paternal warmth, he never indicated overtly that he felt a lack. Later he would complain with a certain bitterness that

*Comrade Van and Comrade Jan Frankel, taking a break at the Dewey Commission hearings. 1937.*

the Old Man did not care deeply about him personally, but perhaps, without realizing it, he had contributed as much as his boss to the "all business attitude."

In terms of the business of politics, Mexico was where van Heijenoort began to feel he had really come into his own. There his confidence grew and he became an expert Marxist–Trotskyist revolutionary-in-exile, as knowledgeable in the art of gaining small favors as in the dialectic. He also continued to be very good at getting along with Trotsky and Natalia, an art in which very few were as skilled as he, for it meant, among other things, keeping unsolicited opinions to oneself. Jan Frankel, for example, who had followed the group to Mexico in February 1937, moved out of the blue house in April because he could no longer put up with Trotsky's moods and temper. After Frankel's move van Heijenoort took his place as the unofficial first secretary.

The distinction of being first secretary did not, unfortunately, relieve Van of his former job as first "gofer," which meant attending to the humdrum and the petty, and learning how to cope in a social culture which, to him, was as exotic as its vegetation. For although the Mexicans were exceedingly kind and friendly, the tempo and patterns of their life and their ways of working were completely different from his European habits, and he found it especially difficult to get organized.

In addition to all the pressures of the moment, just as the exiles arrived in Mexico, the second of the purge trials began in Moscow and the sensational news echoed around the world. The Mexican press, as well as the international press, carried daily headlines and stories about the accusations against Trotsky and his son, Leon (Liova) Sedov. No longer gagged, as he had been in Norway, Trotsky immediately launched a defense and began writing his own account of what was going on in Moscow. In *With Trotsky in Exile,* van Heijenoort describes the excitement:

> Each day, news releases arrived reporting the charges fabricated in Moscow and each day, Trotsky responded with articles exposing the mechanics of the fraud. Each article had to be immediately translated into English and Spanish, then distributed to the international news services and the Mexican newspapers. In the evenings I made the rounds of the editorial offices of Mexico City's dailies, giving them Trotsky's statement of the day.

The international press took up the controversy, and stories and pictures of Trotsky and van Heijenoort appeared in the newspapers of all the major capitals of the world. Thus, one day, out of the blue, van Heijenoort received a letter from Delft, Holland, written in Dutch, which—his name notwithstanding—he could not read. Eventually he found a translator and to his amazement learned that the letter was from a woman who had seen his name and picture in her newspaper. She was his

*The philosopher John Dewey, head of the Commision of Inquiry into the Moscow Trials. Coyoacán, 1937.*

aunt, she said, and his father was her brother. The name van Heijenoort was rare, even in Holland, and the basic facts she gave corresponded with the little Van knew about his father. Through this contact, van Heijenoort was eventually to learn more about his paternal ancestors in the Netherlands than he had previously known.

Trotsky, in his defense against Stalin's attack, wanted to do more than engage in journalistic arguments; to this end, he asked "for the formation of an international inquiry commission that would examine the false accusations made against him and his son in the Moscow trials." He wanted the hearings of this board to be the equivalent of a trial where in response to questions by the examiners, he would have the opportunity to deny the charges against him and Liova. And, not incidentally, he hoped his "trial" would offer him a platform from which he could denounce Stalin, Stalinists, and Stalinism—in court, as it were.

Within a few months the accused man got what he wanted. An international commission, consisting of six Americans, one Frenchman, an Italian, and a Mexican, was formed, and, to the Trotskyites' great satisfaction, John Dewey, the eminent American philosopher, accepted the job of chairman. Dewey was a liberal but by no means a Marxist, and any judgment handed down by a committee he chaired would carry weight because of his great stature as a fair-minded, disinterested, but responsible thinker. According to Professor Sidney Hook, who was Dewey's student and later a close colleague, there had been heavy pressure from various quarters to convince the distinguished man to refuse. His family feared for his health and safety in Mexico and pleaded on those grounds. (Dewey was seventy-eight at the time of the hearings, but he was to live another fifteen years.) And, the greater majority of the radical-left, who supported Stalin to one degree or another, also urged Dewey to decline. If he had wavered before, these attempts to stifle the hearing convinced Dewey that he had to do the job.

The establishment of the board of inquiry, which came to be known as the Dewey Commission, meant that the group in Coyoacán had to get to work immediately to prepare for the inquiry. Masses of documents had to be read, including the papers Trotsky had taken with him when he had been banished from Moscow in 1928 that no one had looked at since. Affidavits about the assertions made at the Moscow trials had to be collected, translated, and annotated in order to be understood by the public and the commission. For weeks there was feverish activity in the house at Coyoacán. Every morning, at a meeting in Trotsky's study, tasks were assigned. Countless minute details had to be clarified, explained, and arranged. To those present it seemed that Trotsky was once again the organizer he had been during the years of the Revolution. Later van Heijenoort would write: "Needless to say, in all this work, there was nothing falsified, nothing hidden, no thumb pressed upon the scales."

It was not just Trotsky who rose to the occasion. Never were Comrade Van's abundant talents for organization and attention

*Van (carrying papki) and Trotsky, during the Dewey Commision hearings. 1937.*

*Leon Trotsky and Diego Rivera. 1937.*

to detail put to greater use. In addition to his intensive work on the documents and the daily press releases, he had the ever-present problem of security, now further complicated by the number of journalists and guests who were going to come to the hearings. In consultation with the Mexican authorities, the group decided that the safest place to hold the inquiry would be in their own blue house in Coyoacán, which, from the moment of their occupancy, they had made secure by replacing windows with adobe bricks, strengthening garden walls, and purchasing additional land as a buffer. All this had been done with the blessing and pocketbook of the generous Diego Rivera. However, certain logistical problems remained, not the least of which was that there was only one toilet to take care of the needs of the forty to fifty extra people who would be present at the week-long "trial."

The hearings were held from April 10 to 17, 1937. Photographs of the proceedings show Comrade Van on Trotsky's right, alert and ready to furnish the needed reference or document. The great effort expended on preparing the evidence and the environment for the hearings paid off. In that controlled but rather intimate setting, Trotsky was simultaneously brilliant, charming, and convincing. He testified six hours a day for a week, answering questions posed by all the members of the commission and providing evidence for each answer; and in closing his testimony, he gave a speech with just the right measure of history, reason, and passion to bring forth a burst of spontaneous applause from the audience. As the sessions were drawing to a close, but well before the decision was handed down, there was this little flourish between Chairman Dewey and Leon Trotsky: "If all Marxists were like you, Mr. Trotsky, I would be a Marxist." To which Mr. Trotsky replied, "If all liberals were like you, Mr. Dewey, I would be a liberal." After it was all over, Dewey wrote to his friend Max Eastman, "You were right about one thing: if it wasn't exactly a 'good time,' it was the most interesting single intellectual experience of my life."

A few months later, the brief romance between Dewey and Trotsky came to an abrupt end. When the commission gave its report and announced its verdict of "not guilty," Chairman Dewey used the occasion to make some personal remarks criticizing Bolshevism, and Trotsky was absolutely furious. Nevertheless, the Old Man had gotten what he was after: a public hearing and an acquittal by this prestigious though non-legal body on the charges of treason. Most importantly, the proceedings and results of his "Mexican trial" had been widely publicized and favorably reported upon all over the world. For van Heijenoort, personally, these positive results were a source of great satisfaction. It was one of the rare occasions where the solution to a problem had come about almost exactly as he and his comrades had hoped.

When he later described the enormous amount of work the Trotskyites had to do to prepare for the hearings, van Heijenoort wrote the odd sentence: "Needless to say, nothing was falsified, nothing was hidden." Why needless to say? Why *wouldn't* Trotsky, the ex-commissar of war, have plotted a return? Why wouldn't there have been *many* plots against Stalin by those who, early on, were keenly and painfully aware of his diabolical cruelty and singlemindedness?

Fifty years after the Moscow trials, van Heijenoort's response to such questions was an excited and emphatic:

> Yes, that's exactly my opinion. There should have been much more. You know there was something very specific on the personal plane about Stalin, something in him, a kind of killing mania that was not in Bukharin or Zinoviev, so it would have made sense to kill Stalin personally. But Trotsky always said 'We are against personal terrorism.' I say that is bunk. Of course Stalin should have been eliminated.
>
> It is true that Kirov did send a man to Paris to contact Trotsky [as it turned out, Kirov paid for this with his life], but Trotsky was not there, so the man was in touch with Liova Sedov instead; and then things moved too fast. It is also true that there were a lot of things brewing in the Soviet Union from 1932 until Kirov was killed in 1934, but Trotsky was never the initiator. He was badly informed, and he really didn't push it anyhow.

It is not within the scope of this work to go into the details of the purge trials or what went on in the Soviet Union in the years that preceded them. Van Heijenoort called it a "fascinating story" and indicated that until Stalin began and persisted in his

*Consultation during the hearings. From left: Van, Albert Goldman (Trotsky's lawyer), Trotsky, Natalia, Jan Frankel. 1937.*

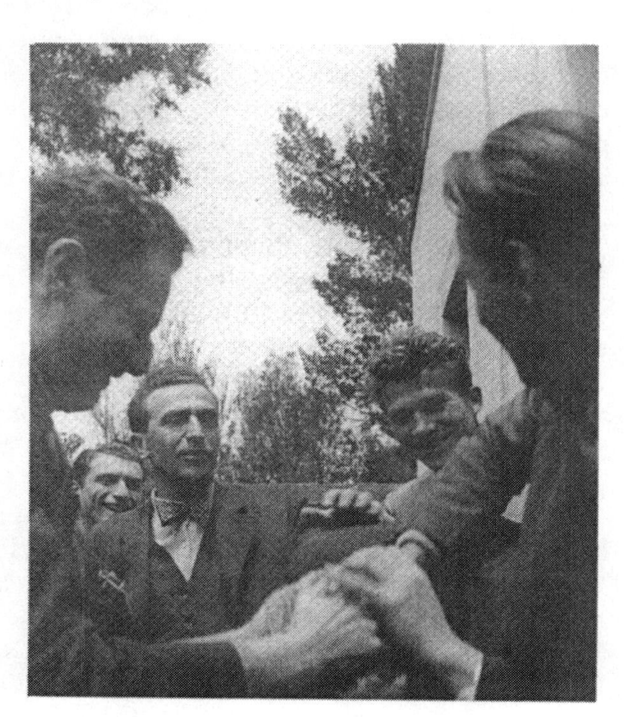

*A light moment during a break. Van, Jan Frankel, and others, thumping Albert Goldman's head. 1937.*

insane, wholesale drive to eliminate all those he perceived as enemies, including his former allies, Trotsky and his comrades thought there was legitimate reason to be optimistic about the possibility for return. What is most interesting in this connection is that Van faulted Trotsky and his followers for having done *too little* in the way of conspiracy.

By the time the group came to Mexico it was certainly too late: there was no way to justify optimism about gaining a power base in the Soviet Union or anywhere else. Instead, the comrades just kept plugging along, doing what they could, mostly by way of trying to influence others. And there were compensatory fringe blessings: Mexico was indeed van Heijenoort's kind of place; he was particularly sensitive to its lush beauty. He loved the warm climate, the local color, the primitive quality of ordinary people's lives. Rousseau's romantic picture of the natural man had influenced him in his secondary-school days, and his appreciation for that ideal had not left him. The world situation may have been dreadful, but right there in Coyoacán one would never have known it. He and his comrades had a comfortable house, plenty of domestic and secretarial help, a contingent of guards supplied by the Mexican police department as well as their own hand-picked Trotskyite guards; finally, they had, as their very good friends, important people with influence.

In a certain sense, their conditions were almost too good. The quality of assistance they were given combined with the natural kindness and warmth of the Mexican people made it extremely difficult to maintain the twenty-four-hour vigilance with the usual rigor everyone thought necessary to protect Trotsky; and while the Cárdenas government was officially nonpartisan, the Trotskyites knew that they had a good deal of official sympathy and support that went beyond the general friendliness with which they were received. In no other country had they been given direct help in dealing with specific problems, but in Mexico, Antonio Hidalgo, one of Cárdenas' friends, "an upright man of great character," was assigned to the group as a liaison. Perhaps Hidalgo was also supposed to make sure that Trotsky kept his promise of non-interference in Mexican affairs, but in any case, he was ready with effective help whenever a difficult matter needed to be settled with a local official.

Above all, they had as their benevolent host and guardian, Diego Rivera, who had been instrumental in getting Trotsky into Mexico in the first place, and had shown unflagging concern for his well-being. The unique friendship that developed between the two great men, both prima donnas, was not without its dangers and disappointments; but for the first year and a half, Rivera and his wife, Frida Kahlo, were truly a pair of ministering angels, Rivera on the political and Kahlo on the domestic side. Rivera had long been a Marxist and a Communist, but in the late 1920s he had come into conflict with the Communist party. Changing his Marxist slant in the early thirties, he became a great admirer of Leon Trotsky.

Frida Kahlo shared her husband's political sympathies but never actually joined the Trotskyite party as he had done. In van Heijenoort's view, she was "not a political person. She just went along with what Diego believed." As domestic angel, however, she helped with the hundreds of details of everyday life. She gave the Trotskys general advice, provided them with servants who could be trusted, and, with her intense beauty, presented a display of free-wheeling style and behavior the likes of which they had never seen. She attended all the Dewey Commission hearings, wearing her brilliant Indian clothing and Tarascan jewelry, always sitting as close to Trotsky as possible. Diego Rivera, of course, was there, too, sporting a wide-brimmed hat with a peacock feather. Frida was small, Diego was gargantuan. They were a spectacular pair.

After the Dewey "trial" was over, contact between the Riveras and Trotskys continued with dinners, excursions, and picnics; and the not-so-subtle flirtation that had been going on between Trotsky and Frida Kahlo during the hearings burst into a full-blown romance. At thirty, Frida was a stunning woman, "remarkable for her beauty, character, and intelligence," van Heijenoort wrote later. "In her relations with Trotsky, she quickly adopted a certain freedom of manner. Frida did not hesitate to use the word 'love,' after the American fashion. 'All my love,' she would say to Trotsky upon leaving." Trotsky found her openly seductive style irresistible. And what did Frida see in the Old Man whom she nicknamed *Piochitas* (Little Goatee)?

*Frida Kahlo. 1937.*

He was heroic, brilliant, charismatic, and, although approaching sixty, vigorous, direct, and quite blatantly amorous.

They began by leaving notes in the books they exchanged, and, in a very short time, for a short period, they became lovers. Although Trotsky was seldom alone (his guards saw to that), he and Frida nevertheless managed to have a few assignations at her sister Cristina's house. Eventually, the love affair became known to everyone in the circle except Diego Rivera.

Natalia, whose whole life and every thought was centered on her devotion to her husband, was jealous, angry, and deeply hurt. She became very depressed. Everyone else in the inner circle felt betrayed in a different way. They were concerned that if scandal leaked out, their enemies, the Stalinists, would surely use it to damage their political cause. They also worried about what Rivera might do if he found out, because although Rivera granted himself freedom to play at love when it pleased him—and it pleased him often—he was not so generous in spirit when it came to his wife's affairs with men. Van Heijenoort described him as "morbidly jealous," and indeed he had threatened to shoot at least one of Frida's other lovers. "He was always threatening someone, and I never knew if he meant it," van Heijenoort later said. "But I didn't say anything to Trotsky, and I didn't pass any moral judgment."

In those days, unless his opinion was asked, van Heijenoort never said anything which could be construed as critical, although his silences were often loaded. In later years, on mathematical or other professional matters or political issues, he was no longer so reticent, but on personal matters he would remain discreet in the extreme and rarely comment upon his friends' romantic arrangements and activities.

Jan Frankel, who had already moved from the house in Coyoacán to an apartment in Mexico City, was less restrained. He took it upon himself to speak to Trotsky about the personal and political dangers of his liaison with Frida. Frankel said what all the other comrades were thinking: that Trotsky was jeopardizing not only himself but also the security of the whole group, and that if Rivera were to discover that he had been betrayed by Comrade Trotsky, the explosion could place their very existence in Mexico at risk. Predictably, the Old Man was furious, and the argument that ensued was the final straw that caused the break in all relations between Trotsky and Frankel. At that point the general level of tension within the inner circle at the blue house in Coyoacán became nearly unbearable.

To escape the pressure at home, Trotsky, with a chauffeur and a bodyguard (not Van), retreated to a hacienda in San Miguel Regla, a town about ninety miles from Mexico City. For three weeks, he stayed in the home of a good friend of Antonio Hidalgo; always ready to help, Hidalgo had arranged even this. Frida came to visit her *Piochitas* at least twice, always accompanied by others, but there the love affair is reported to have ended.

"The stakes were too high. The two partners drew back," van Heijenoort later said, giving the impression that the break was by reluctant mutual agreement. However, Hayden Herrera in her biography of Frida Kahlo suggests that Frida had had enough and found it easy to let go, whereas Trotsky pleaded for a continuation of their relationship. In a letter to her friend, Ella Wolfe, Frida confided, "I am very tired of the Old Man."

Chapter Eight

# Arrivals and Departures

Mexico: 1938

*Trotsky, Diego Rivera, Natalia, Reba Hansen, André Breton, Frida Kahlo, Jean van Heijenoort (in shadow). Chapultepec Park, Mexico City. 1938.*

It was so tense, our small group, Trotsky, Natalia, me; Diego, Frida, Breton, Jacqueline ... so many subtle things going on.

JEAN VAN HEIJENOORT. *1984*

AFTER THE DRAMA of Trotsky's affair with Frida Kahlo, Comrade Van had to admit, however reluctantly, that he now saw the Old Man in a different light. The halo had taken on some tarnish. Of course this was not the first change in his perception. After four years of close and constant association he had accepted the fact that the man he revered had human failings. He had already been well aware of the old man's lively interest in women, particularly Frida and her beautiful sister Cristina but, he said later, "I made no moral judgments about that." His attitude changed when he realized that there was a qualitative difference in this case where much more than romance was involved. By putting his personal emotions before the political good of the group, Trotsky had seemed ready to risk the life of their whole enterprise. Comrade Van would never have permitted himself such indulgence. When the Trotsky-Kahlo romance ended without fireworks and, apparently, without Diego's knowledge, van Heijenoort and everyone else in the circle breathed a great sigh of relief.

In the days and weeks that followed, relative calm was reestablished in the blue house. The dust from the move had settled, the Dewey Commission hearings were over, the Trotsky-Frida crisis had been weathered, and the Old Man and his loyal wife were reconciled. It looked as if the exiles would be staying in Mexico for the foreseeable future—as if, indeed, there was a future. Van Heijenoort had time to think of himself and his family, and he wondered whether he might not ask Gaby and his son to come to Mexico. It did not seem too frivolous an idea, especially in light of Lev Davidovich's recent escapade. After consulting with Trotsky and receiving his consent, van Heijenoort wrote to his wife, said how much he missed her, described the beauties of the place, and urged that she and Jeannot join the group in the blue house in Coyoacán. There

*Gabrielle Brausch and Jeannot van Heijenoort. 1937.*

was one condition: she had to put aside her differences with Trotsky or, at the very least, not express them. After ten months of separation Gaby wanted to be with her husband, so although she was still allied politically with the Raymond Molinier faction in Paris, she agreed to hold her tongue. Van, for his part, in his intense desire to be reunited with his wife, forgot how difficult if not impossible it would be for the forthright Gaby to keep that promise.

To everyone's great happiness, in November of 1937, Gabrielle Brausch and Jeannot, who was then two years old, sailed into Veracruz harbor. Of course, the eager husband and father was there to meet the ship. Forty-six years later, Jeannot recalled the moment:

> I remember my father coming on board to greet me, but, in fact, I don't so much remember my father and my mother at that moment as I do a man on the boat who was a ventriloquist

> and could throw his voice. As for the house in Coyoacán, I do remember two or three things: there were guns everywhere, in every room. One day I picked up a gun and walked into the dining room, and my mother just went crazy.

At first things went quite well with all the van Heijenoorts and the Trotskys living together. There were pleasant meals in the dining room and Jeannot's favorite beverage, *Ovomaltine* (Ovaltine), for breakfast. Gaby helped Natalia run the household as she had done in Barbizon, and if anything troubled her, she kept it to herself. "She fit in," was Van's way of saying, later, that his wife did what he had hoped she would do.

The honeymoon lasted for no more than a few weeks and was supplanted by a simmering tension between the two women which one day came to a full boil, in the kitchen. They were preparing the midday meal with the help of a young Mexican maid who was part of the staff. Natalia, who knew almost no Spanish, gave her instructions with large gestures and a few shouted words and pulled the maid by the arm to direct her attention to the matter at hand. Gaby, already angered by what she considered to be Natalia's uncomradely and imperious attitude towards the Mexican help, seized upon the moment to voice her disapproval. Using frank, Belleville (Gaby's working-class neighborhood) language, she told Natalia what she thought of her behavior. Then, having begun, she realized she had a few more charges to make and permitted herself to do so. The floodgates opened and a torrent of resentment spilled out. Comrade Natalia, a strong woman, was not going to drown in silence; instinctively she moved to save herself and to counterattack.

At the height of the brawl, Trotsky, who had been in his study upstairs, chanced to be on the way to the bathroom downstairs. As he passed the kitchen, he heard the tumult, and went in to investigate. And when Lev Davidovich found Natalia and Gaby screaming insults at each other and near to blows, he himself began to shout, louder than either of them, "I am going to call the police." His cry drew everyone else in the house into the kitchen, and the women were separated and silenced.

The outburst had both an immediate and a very serious long-term effect upon the lives of the van Heijenoorts. Although, according to Van, "nothing more was said," after the intensity of the fight, it was understood by everyone concerned that Gaby could no longer stay in the blue house. After such an incident in that small, close group, it was, in van Heijenoort's opinion, impossible. Gaby concurred. A very Parisian, very Belleville woman, accustomed to speaking her mind, she realized she could never be content as second cook and housekeeper in Natalia Sedova's domain even under the best of circumstances. Even so, she was angry and embittered: first, as a matter of principle, by Natalia's behavior; second, by Trotsky's role in the argument; third, and perhaps most of all, by her husband's passive attitude and lack of direct support for her. Had she traveled all the way from Paris for this?

Within little more than a week after the calamity in the kitchen, Gaby and Jeannot were ready to leave, and, in spite of van Heijenoort's later statement that "there was not much explanation," enough was said, or not said, to make it clear to him that Gaby's departure was going to mean the end of their marriage. Yet he did not dissuade her, nor did he suggest that she and Jeannot might live independently and that he would divide his time, as he had before, between Trotsky and her. What had been true before was still true: whatever the issue, Comrade Van would sacrifice any relationship in favor of his love for Trotsky and what he represented.

Describing his wife's and son's departure, later, van Heijenoort said,

> It was very sad when Gaby left, because it was not like the year before when I left for Mexico, and we didn't know what would happen. This time we both knew that it was really the end. I remember the scene very vividly: it was evening; we went to the big station in Mexico City; I put them on the train, and I wondered when I would see my son again.

The ever-generous Diego Rivera provided the funds for travel, and Gaby and Jeannot returned to France. Again the little boy went to stay with his grandmother Hélène in Creil, and Gaby found a job as a bookkeeper and took up where she had left off with the activists in the Molinier faction in Paris.

But it was Gaby whom Van would never see again, while within a few years, Jeannot would be crossing the Atlantic once more.

To van Heijenoort the most incredible part of the whole affair was Trotsky's behavior. Taking Trotsky's initial words literally, he wondered, rhetorically, what Trotsky thought the Mexican police would do. He knew Trotsky had no intention of calling the police, and that the words were a reflex action, the kind of threat he might make to a child. But that the Old Man could have even said such a thing was very disturbing to Comrade Van, a man who was very careful with language. It was one of the few emotionally loaded personal situations between them that he wrote about directly in *With Trotsky in Exile*:

> On the afternoon of the same day, Trotsky came to see me. "I should not have said that," he admitted in a confused, almost sheepish manner. This was the only time that I ever saw him so embarrassed.

Decades later, van Heijenoort was still trying to analyze the episode. He would tell the story as an illustration of the foolish, irrational side of Trotsky: "Absurd," he would say, "What would a Mexican policeman do?" He also cited it as an example of Trotsky's inability to draw back and listen to other people's explanations instead of jumping to his own conclusions:

> He could have asked: "What's going on here?" but he didn't. It was not in his character. It was the same thing later, in the break with Diego. Trotsky couldn't say: All right, I am going to sit down and work things out with Diego. He was not that kind of man.

Van Heijenoort knew that Trotsky's intervention in the fight between the women had been destructive, that he had aggravated a situation that might have been resolved. Nevertheless, for better or worse, he felt he had to stick with Lev Davidovich without criticizing his actions. Explaining his position, he later said, "If I didn't maintain that uncritical attitude, I knew I would have to leave, and I couldn't do that because there was no one who could take my place."

Duty, commitment, and attachment. Van Heijenoort would always emphasize those feelings when he described why, in spite of Comrade Trotsky's diminished luster, his own loyalty remained staunch. He was attached to "that man" with a bond stronger than the one which had attached him to wife and child. Certainly it was true that Trotsky needed him, and certainly no one had Van's expertise in every detail of working with and "minding" the Old Man. In that sense he was right; no one *could* take his place. But there was another side to the equation that Van never mentioned: his *own* need for the Old Man. If he were to leave Lev Davidovich, there was no one who could take the Old Man's place in Van's heart and mind. What kind of activity could he ever find to replace the mixture of devotion, dependency, excitement, and responsibility that his life with Trotsky demanded? On a rational level he had already had more than a few intimations that he was pursuing a lost cause, but whether he knew it or not, Comrade Van needed Comrade Trotsky as much as or more than Trotsky needed him.

In February of 1938, not long after Gaby's departure, news came from Paris that Liova Sedov had died in mysterious circumstances following what appeared to have been a successful appendicitis operation. Encouraged by Mark Zborowski, a Russian "friend," Liova had chosen to go to a hospital run by Russians instead of a French hospital. He did not ask the advice of the leaders of the Trotskyite group because, with a few exceptions, he did not like or trust his French comrades. (Van was one of the exceptions.) This lack of confidence may have cost him his life. Gérard Rosenthal, whose father was a prominent Paris doctor, could have insured Liova of excellent care had he known of his illness. Instead, without informing anyone except Jeanne Martin and Zborowski, Liova ensconced himself in a Russian clinic, which in Paris in 1938 could only have been staffed with White Russians and Stalinist agents, both dangerous adversaries. "He adopted the ridiculously transparent pose of being a French engineer," van Heijenoort later said. "In two minutes, other Russians could not have failed to realize that he was Russian."

Trotsky and Natalia were grief-stricken and secluded themselves for days. Without knowing the full details (they would

*Gaby and Jeannot (lower right) sailing to Mexico. 1937.*

never be known) they were convinced that Liova had been murdered, and their suspicions were not without grounds. Mark Zborowski, Liova's "friend," was later revealed to be an agent of the Soviet secret police. In the 1930s, while he was studying anthropology at the Sorbonne, he befriended Russian dissidents and passed names and information to the GPU. In 1940 he came to the United States where he continued his activities as a spy. He was convicted of perjury and spent time in prison in the 1950s. After his release, in an apparent turnabout, Zborowski moved to San Francisco and established a Pain Center at Mount

Zion Hospital where he did research into the cultural aspects of pain.

Liova's death compounded van Heijenoort's feeling that now, with such sadness and torment in the household, there was no way he could leave his leader. Within weeks of his son's death, Trotsky was quarreling over Liova's papers, which he insisted should be sent to him because they were legally and morally his. But Jeanne Martin, Liova's companion, who still belonged to the dissident Molinier faction, refused. While Trotsky worried that the French police or the GPU would get hold of the documents (in fact, some of the papers had already been stolen by the GPU prior to Liova's death, and Zborowski, it seems, was also involved in the burglary), Jeanne Martin and Natalia wrote sorrowful but friendly letters of consolation to one another. This correspondence further infuriated Trotsky, and he accused his loyal wife of siding with his enemies.

The gloom in Coyoacán was relieved to some extent by the arrival of André Breton, who, sponsored by the French Ministry of Foreign Affairs, had come to give some lectures. His presence in Mexico as the founder and chief exponent of the surrealist movement created a great deal of excitement and was the occasion for many meetings, banquets, and parties, official and unofficial. Breton was not one of the Trotskyite activists, but he was a well-known sympathizer. He had worked on Trotsky's behalf in Paris in 1936 when the commission of inquiry on the Moscow trial had been formed there; and he had had bitter fights with his former friend and comrade-in-arms Louis Aragon, the surrealist poet who had embraced the Stalinist position on revolution.

As soon as Breton and his wife, Jacqueline, arrived, van Heijenoort, as Trotsky's emissary, went to greet them and to offer his services as a translator and guide. Breton had retained a very favorable impression of Comrade Van from their previous brief meetings in Paris. He accepted the offer with pleasure, and van Heijenoort, who had admired the surrealist from afar, was elated to have him at close range. He showed the Bretons the sights of Mexico City and took them to the most typical restaurants. The Bretons were enchanted by everything they saw and everyone they met.

Van Heijenoort was surprised by Breton's atypical effusiveness. The autocratic *chef du corps* of the surrealists, who, in Paris, decided who was "in" and who was "out" and ran his group like a political organization, was behaving like any other tourist under the spell of an exotic culture. "Everything in Mexico was so new to them," Van said later, "the mangoes, the papaya, all the tropical fruits; the flowers, the people, Frida, Diego. 'The surrealist place *par excellence*,' was what Breton called it. He even liked the Mexican poets!"

The Bretons stayed for several months, living part of the time as guests in the Rivera house in San Angel. Very soon after their arrival, van Heijenoort brought them to Coyoacán to meet Trotsky and Natalia. Van, who was present at this as well as their subsequent meetings, took notes on their conversations, which ranged from discussion of the work of the Paris commission of inquiry on the Moscow trials, to a comparison of literary realism as exemplified by Zola, to surrealism, to Freud, and to the role of art in society. Comrade Van soon thought of himself as the *"trait d'union"* (hyphen) that connected Breton and Trotsky, a role he enjoyed because of his strong identification with both men and their ideas. There were, however, some exceptions to this pleasure, and on several occasions, when he was caught in the cross-fire between his two heroes, he found it necessary to do some very fancy skipping and dodging. For, in spite of the great respect they had for one another, predictably the surrealist and the Marxist-realist came into conflict, and when they did, Van tried his best to avoid having to choose sides.

Although Breton and Trotsky were in accord on most political issues, they were strikingly different in temperament and in the manner in which they focused their attention, and this led to many disagreements about the purpose of art. In their early discussions they argued about the unconscious, which Trotsky recognized as important, but considered as a realm to be penetrated and conquered with reason. He accused Breton, who was an early advocate of Freud's theories, of cultivating the unconscious for its own sake. Breton maintained that the unconscious and the fantastic could not and should not be reduced to reason; these things must be allowed to exist alongside, if not against, rational thinking. At that early point in their friendship, both men were careful not to push their arguments too far. Instead, they sought and found many subjects upon which they could agree.

If Mexico, for Breton, was "the surrealist place *par excellence,*" in Frida Kahlo he found the surrealist artist to represent the territory. He labeled her art: "a ribbon around a bomb."

Captivated by the land and the people, as well as by his own interpretations of Mexican life and the country, Breton was avid to see and experience as much as possible, and everyone was eager to please him and his wife. Grand excursions and group picnics were organized with Frida, Diego, Trotsky, Natalia, van Heijenoort, and other secretaries, guards, and chauffeurs. In a caravan of two or three cars, the group would go out into the countryside, to the towns and villages near Mexico City. There, while taking in the local culture, they would embark upon long discussions about life, art, and politics. For a while Trotsky spoke seriously of publishing some of these conversations, but in Patzcuaro, where the idea was beginning to take a more concrete shape, Breton had a sudden—perhaps convenient—attack of aphasia, and the plan, it seems, also lost its voice.

Trotsky had another idea. Almost as soon as Breton had arrived, Trotsky proposed that they write a manifesto calling for the creation of an International Federation of Independent Revolutionary Artists. The operative word here was *independent* revolutionary as opposed to Stalinist revolutionary, for, as in every other area, the Trotskyites wanted to make their position distinct. Breton and Rivera agreed to collaborate, and Breton quickly accepted the responsibility for writing a draft of the proposed manifesto. As it turned out, for Breton, agreeing was one thing, but doing quite another. Van Heijenoort later gave this account:

> Soon afterward, Trotsky began to press Breton for the draft of the manifesto. Breton, with Trotsky breathing down the back of his neck, felt completely paralyzed, unable to write, which was quite understandable. "Have you something to show me?" Trotsky would ask whenever they met. As the situation developed, Trotsky assumed the role of the schoolmaster before a Breton playing the recalcitrant pupil who had not done his homework. Breton was

keenly embarrassed; the situation dragged on. One day in the garden at Diego Rivera's house, he took me aside to ask me, "Why don't you write this manifesto?" I declined, not wanting to muddle the affair still more.

Time passed and still Breton had nothing to show. One day, while he and Trotsky were together in a car, on one of the groups' outings to Guadalajara, the Old Man got so angry at the surrealist that he ordered his driver to stop and told Breton to get out and ride in the other car, which was following behind with Van, Frida, and Jacqueline. Van Heijenoort walked forward to see what was the matter, and to him Trotsky said, "Get in." As they passed to trade places, Breton made a gesture of baffled astonishment. Van got into the first car and they started off. Trotsky remained in the back, upright and silent. He gave no explanation of what had happened. Nor did Breton ever explain what particular spark had fueled Trotsky's anger, although at the time it seemed clear that it had something to do with the nonwriting of the manifesto. It could also have been that the Old Man was annoyed and jealous that Comrade Van and Frida, in whose friendship he sensed an increasing intensity, were riding together in the second car.

Eventually, Trotsky's wrath subsided, and Breton managed to produce a first draft which he gave to Trotsky, who then wrote a few pages of his own. Van Heijenoort translated them into French, and they were added to Breton's text. When the manifesto was finally finished, it was signed by Breton, Rivera, and Trotsky. Rivera had not written a single line, but he had had some ideological input, and it was reasoned that since the document spoke directly to independent revolutionary artists, his signature enhanced its value. Ultimately, the manifesto was published in *Partisan Review*, then translated into several languages, and widely read in radical intellectual artistic circles.

The farewell meeting between Trotsky and Breton took place at the end of July 1938, on the sun-drenched patio of the blue house in Coyoacán, amidst cactus, orange trees, bougainvilleas, and pre-Columbian statues. Trotsky was in an expansive mood. Making what seemed to be an impulsive decision, he went to his study and returned carrying the original joint manuscript of the manifesto. In an extremely unusual gesture, something van Heijenoort never saw repeated in all his time with Trotsky, the Old Man presented it to Breton. The men parted on good terms and Comrade "Trait d'union" could relax for the moment—but *only* for the moment, because between Trotsky and his artist friends, matters were always on the edge. Here is van Heijenoort's recollection of his feelings:

> It was so tense, our small group, Trotsky, Natalia, me; Diego, Frida, Breton, Jacqueline ... so many subtle things going on. Of course, there really was a lot of good will too, and we did have the feeling that we were going to create something special: a fusion of politics and art that we would use to oppose to Stalinism in all its forms. We all had a great desire for that, but in practice it just didn't come off at all, and then finally there was a split, with the politicians on one side and the artists on the other.

The "subtle things going on" to which Van alluded were more significant than anyone admitted. These Marxist men of reason played down the emotions and sometimes successfully kept them down. But inevitably occasions arose when the powerful vapors of jealousy and irrational love or hate did bubble up and explode right in the face of the dialectic.

When Breton left Mexico, the split between the artists and politicians had not yet occurred. Upon his return to France he gave a stirring speech describing the wonders of his stay in Mexico. In the course of this talk, Breton took time to praise "le Camarade Van" placing his panegyric immediately after one of his denunciations of Louis Aragon. After calling his former friend a liar, a hereditary spy, and a traitor, he had this to say of van Heijenoort:

> The day after my arrival, I was overjoyed to meet Comrade Van, whom many of you know. Anyone who has had anything to do with him is aware of his extraordinary intelligence and sensitivity and the quickness and clarity of his judgment, but not

everyone has had the opportunity to appreciate the breadth of his interests and the goodness of his heart. He is so modest that he will certainly be offended by my words, but I would be angry with myself if I did not take this occasion to give him my truly brotherly praise. I hope he will pardon me for speaking about how moving his life is, but one has to cite him as a positive example in opposition to all those intellectuals who seek a comfortable life while denying all moral conscience.

At the age of eighteen Comrade Van, who had been admitted to the École Normale Supérieure, was not able to bear the thought of Comrade Trotsky's isolation in Prinkipo, and giving up his own future, he spontaneously offered his services to Trotsky. He followed him everywhere in his exile, passing through the hands of almost all the police of Europe. At present, he is very poor, because Trotsky does not have the means to give anything to his secretaries except room and board. He continues to live without having the least little thing in the way of personal possessions. He is even deprived of the smile of his child.

It is with all the good will in the world that he takes on his overwhelming task: ten to twelve hours a day of work and then, since there has to be continuous surveillance of the house, four hours of guard duty each night. Comrade Van is a revolutionary from head to toe—everything Trotsky could want in a man.

That night [Breton is referring to their first dinner together in Mexico] in the relaxed atmosphere of the restaurant, while the waitresses in their beautifully embroidered Tehuantepec blouses were moving around us, he answered my flood of questions with his beautiful smile. To console us for so many others, here is a real man, a friend in every sense of the word.

How could the subject of such praise from a great man fail to be flattered? Yet, at the time, van Heijenoort was less pleased than embarrassed by the excess of it all, and he was disturbed by the inaccuracies of detail. He went so far as to write Breton to tell him of his embarrassment. He had been twenty, not eighteen when he had gone to join Trotsky, and it had not been his "spontaneous idea to offer his services;" he had been asked. But most distressing was the description of him as "very poor." He found Breton's choice of words worse than inaccurate because, as he would say later, "it totally missed

*The poet, the painter, and the revolutionary. Coyoacán, 1938.*

the point." As far as Comrade Van was concerned, his poverty was importantly unimportant. Furthermore, he worried a great deal about what Trotsky might think about Breton's statement. Would the Old Man think that he had complained to Breton about his situation with Trotsky, about his pay, or rather his nonpay, and the fact that he had no money that he could call his own?

Van Heijenoort insisted that it would have been inconceivable for him to make a complaint of that nature because money played absolutely no part in motivating the choice he had made. He had come to serve a cause; he was a comrade. Did the disciples expect payment from Jesus? He felt no more deprived than they. Breton's misconception of Comrade Van's attitude was a clear sign that the surrealists and Trotskyites were living in different worlds that had as much to do with personal style as with political vision.

Still, although Van may have found Breton wanting in his perceptions of the workings of the inner circle, he liked him very much and had had no problems in getting along with him.

He responded to his effusive, open nature and would continue to have what he called "warm relations" with him for many years. "We were good friends; we were easy together." In the few letters he wrote to Breton, there is a lighter tone than in those he wrote to Trotsky.

The calm that followed Breton's July 1938 departure was interrupted in December when Diego Rivera asked Van to help him write a letter to Breton. This letter would lead to open conflict and animosity between Rivera and Trotsky, and once more Comrade Van would be forced to choose between two people he loved, admired, and respected. Except for the distressing episode in the kitchen that had precipitated his separation from Gaby, the rupture between Trotsky and Rivera would cause Van more pain than any other break between the comrades.

Diego Rivera had monumental talent, energy, and drive. Constant and unswerving in his commitment to art, he was otherwise a fickle man, in the habit of indulging in enthusiastic but whimsical swings of attitude regarding politics as well as other more personal activities. One day he would say he wanted to take a more active role in the leadership in the Trotskyite party and the next he would say he wanted to resign from the Trotskyite group altogether and devote himself exclusively to painting. Trotsky, who didn't want him to do either of those things, thought he had convinced him that he could best serve the cause through his art.

Rivera frequently referred to himself as "a bit of an anarchist," an aspect of his character that was very hard for the orderly, organized Trotsky to deal with. As they got to know each other better, the politician and the painter began to find it difficult to keep from quarreling. Trotsky thought he knew about art and its purposes, and Rivera thought he knew about politics. Disregarding his promise to President Cárdenas, Trotsky was, in fact, exerting influence in Mexican politics, especially in the Trotskyite party, and his authoritarian stance was irritating to the irrepressible Rivera, particularly when the Old Man advised him to stick to his art and stay out of party administration.

Breton's heady presence in Mexico had increased the difficulty, and the air of confrontation that developed along the lines of art versus politics did not disappear when he returned to France. It was not that Rivera or Breton ever advocated the doctrine of art for art's sake, but neither did they believe that all art had to be politically explicit. After Breton's departure, Rivera's annoyance with Trotsky did not subside. When Trotsky influenced an editorial appointment for *Clave,* a new Marxist journal, Rivera called it a "friendly and tender coup d'état."

Aggravated by this incident and another where he thought the Old Man had overstepped bounds, he wrote a letter to Breton criticizing Trotsky's "methods." Since the letter was to be in French, Rivera asked for Van's help, and, not knowing Diego's intention, Van agreed. Diego dictated and Van wrote, but when he came to the critical part of the letter he stopped and looked at Diego, saying in effect, you know I can't do this. Rivera urged him on saying, "Write! Write! I myself will show the letter to L.D." (Rivera and others often called Trotsky L.D., and Trotsky often signed his letters with those initials.)

Had it been anyone but Diego, van Heijenoort said later, he would have walked out at that point, but relations between Trotsky and Rivera were exceptional. Rivera was the only person who could come to the house at any time without previous arrangement. Trotsky always received him warmly and met with him privately; with other visitors, there was always a third person present, most often van Heijenoort. According to van Heijenoort, their relationship was "a privileged sanctuary that escaped the system, and I made a deliberate attempt to keep out of their conversations." So he accepted Rivera's promise in good faith, preferring that Diego explain himself directly to Trotsky.

Van and Rivera finished the letter. Returning to his room in the blue house, Van typed it and left the copy on his own desk. Natalia came into his room, found the letter and took it to Trotsky, who read it and, predictably, had a fit. Saying that the accusations in the letter were not true, Trotsky immediately dispatched Van to Rivera with orders to ask him to change the letter. Rivera promised he would do so, but in the end he sent the letter unchanged. The argument between Diego and the Old Man grew bitter. What had begun as something like a minor personal complaint, albeit in a political context,

*Van, Carlotta Fernández (a comrade), Trotsky, Natalia, and Frida's arms and legs. Natalia cut the rest of Frida and all of Diego out of the photo after rupture between Diego and Trotsky. 1938.*

escalated into more serious general differences. A few weeks later, Rivera supported a presidential candidate whom Trotsky opposed. Construing this act as political betrayal, the Old Man directly accused his dearest friend. It was the absolute end. Rivera resigned from the Fourth International; Trotsky said publicly that he no longer felt any "moral solidarity" with Rivera; and the two men never saw or spoke to each other again.

The question of what role Van played in the final split presents itself. Since he had to have been aware of the reaction it would provoke, why did he leave Rivera's letter where Natalia would see it and pass it on to Lev Davidovich before Rivera himself had a chance to show it to Trotsky? True, van Heijenoort always put his letters where they could be seen, but this letter was clearly in a different category in that it was *Diego's* letter to Breton. Did he derive a certain satisfaction from observing others speak or write negatively about Trotsky—a luxury he could not permit himself? "I don't think that's what I was doing," he said years later, "at least not consciously." If the ground here seems slippery, the suggestion feels solid. After all, Comrade Van had reason enough to want to get back at his Old Man now and then.

By early 1939, when all of this came to a head, van Heijenoort was far from being the wide-eyed disciple he had been in earlier years, taking Trotsky's every word and attitude as gospel. Still, although he may have taken vicarious pleasure in Diego's thrusts, when Trotsky decided Diego was an enemy, Van's overriding loyalty to the Old Man compelled him to forsake his friendship with the painter. He was very much upset by the break because he admired and had great affection for Diego, for his great kindness and generosity, and also because he was an extraordinary character—dramatic, colorful, and unpredictable. And there was one very important but secret personal reason for Comrade Van's dismay: if the friendship and collaboration with Diego Rivera was ended and the Old Man and the painter were at war, there could be no separate peace with Frida, and this detail made Van unhappy in a way he could reveal to no one.

Chapter Nine

# Two Women

Mexico: 1938–1939

*Frida Kahlo. 1939.*

Ah, Frida. She was one of the great women of my life. Deeply sensual, extremely intelligent, strikingly beautiful, she was like no other woman I have ever known.

JEAN VAN HEIJENOORT, *1984*

NEARLY A HALF-CENTURY after he last saw her and thirty years after her death, Frida Kahlo's effect upon Jean van Heijenoort was still evident. Whenever he described her appearance and her personality, his mood changed. Physically his body seemed to grow lighter and lift. His ingrained discretion and good manners prevented him from speaking directly about his brief love affair with her, but everything else in his attitude bespoke the hope that the listener would understand that his relationship with Frida went far beyond ordinary friendship.

Seductive, sensual, yet at the same time matter-of-fact and earthy, Frida, of course, seduced him. He would not have dared to initiate the affair. The dangers were all too obvious, and although he was much in the habit of taking risks, to hazard the jealousy and anger of both Trotsky and Diego was out of his line. But once Frida made the move, it was impossible for him to resist. If there was something vaguely Oedipal about his relationship with Frida—after all, she had been his Old Man's lover—it does not appear that he thought of her as "motherly." Instead, she was more like a buddy, a *cuate* or *cuatezone* (a pal or big pal), as Frida would have said.

She was thirty-one, five years older than he, but he called her "boyish, like an ephebe." He thought her masculine ways enhanced rather than detracted from her sexual appeal. She was enchanting, with her odd mixture of feminine warmth and concern, her deliberately vulgar language, her free spirit and devil-may-care flouting of convention. She had a loose tongue, a sarcastic and penetrating wit, and no inhibitions. She shocked Comrade Van and reminded him of what it meant to be free, and he loved every bit of it. "Frida ... she would say *anything*, and so, you could tell *her* anything. She used to say her idea of a good way to live was to make love, take a bath, and make love again."

Frida confided in Van in a way that no other woman had ever done: she told him about her other lovers of both sexes, about her affair with Trotsky that had created such anxiety among his followers, about her earlier liaison with the sculptor Isamu Noguchi, whom Diego had threatened with a gun, and on and on. Van listened to all of this and accepted her as she was; he never felt jealous or possessive about Frida, which for him was most unusual. He even discounted and forgave her later turn to Stalin, a latitude he offered almost no one else.

From the moment of his arrival in Mexico in January 1937 until Frida's departure in October 1938 for the exhibitions of her work in New York and Paris, van Heijenoort and Frida saw one another nearly every day. However, the brief period when they were physically intimate did not take place until a year after Frida's affair with Trotsky had ended and six months after Gaby and Jeannot's departure. It coincided with and was abetted, in a practical way, by André Breton's visit in the spring of 1938. During that time there was a shift in the normal day-to-day routine of the Trotsky household, and Comrade Van had more opportunity for unusual activities, such as lovemaking. Even so, the occasions were few since the obvious constraints and dangers were many.

As far as van Heijenoort knew, neither Trotsky nor Rivera ever found out about his relationship with Frida, although they were aware that he saw her all the time. Here again Comrade Van's innocent look was in his favor. But if Trotsky wasn't specifically jealous or suspicious, he was angry on the rare occasions when Comrade Van went out "on the town." "Trotsky did not like me to be out of the house at night," Van recalled. "He would be mad; I could feel it. Even if there was nothing that I had to do, he wanted me there. But I went anyway. I needed some kind of release from the tension."

It was one of his few acts of filial rebellion. Whether his Old Man liked it or not, Van would go with Frida and her beautiful sister Cristina to El Salon Mexico, the famous dance hall. There, in an enormous L-shaped ballroom, different orchestras would play at the two ends of the room, making no attempt to coordinate their selections. At each end of the hall, people danced normally to the music of the group they were close to and could hear, while at the bend of the L the dancers, hearing both bands at once, would do all kinds of wildly inventive steps to accommodate the cacophony.

Van liked taking part in what he called "the popular night-life activity," and he packed as much action as he could into these rare sybaritic expeditions. He and the Kahlo sisters would join the crowds in the street and make the rounds, going to several night spots. Among his favorites was Tenampa, an after-hours joint where the Mariachi bands came to unwind and play for each other after they had finished playing in the regular clubs.

"At Tenampa, they served only one thing, something they called 'punch,' a thing that could kill you," van Heijenoort gleefully recounted. "They didn't ask you what you wanted. They just gave you the *punch.* People began coming in to Tenampa about 4:00 A.M. At six, when the sun was rising, the musicians were in full swing and at eight, when it was full daylight, everyone went home. I really liked that place."

When André Breton, after a few days in Mexico, pegged it as "the surrealist place, *par excellence*," van Heijenoort, in deference, did not disagree out loud. But later, in spite of his great admiration and affection for Breton, he resented and took exception to his naming. "The surrealists just wanted to claim Mexico and Frida for themselves," he said, "to act like *they* discovered it and her," and, in a rare lapse of decorum, he added, "That's crap!" Most of all, he was angered by the de-personalization of Frida. To him, she was unique, an original, an exotic bird in a tropical land, and the attempt to make her part of an intellectual movement offended him. When he described her as "different," he was referring to her inner qualities: the mixture of kindness and toughness, her frank sexuality, her ability to handle the difficult situations she got herself into. He appreciated her judgments and insights about people, which he thought were "razor sharp," and he, a practiced master at holding back feelings, marveled at the free and easy Frida, who could be brutally frank when someone said or did something she thought was foolish.

All this focus on the inner Frida and her character traits does not mean that Van was oblivious to her outward appearance; far from it. When asked, he would say that he *loved* the way she dressed, a rare use of the word in that context, for usually he saved his "love" for deeper attractions; but he did love the effect Frida created when she dressed in her Tehuana Indian dresses and adorned herself with native earrings and jewelry. The whole production thrilled him. And, although van Heijenoort objected to Breton's labeling of Frida as a "surrealist artist," he thought his description of her work: "a ribbon around a bomb," was absolutely on the mark. Van Heijenoort had made his own assessment of the quality of Frida's work long before she had any real stature outside the small circle of artists in Mexico: "Her paintings were small, her productivity was moderate, and her fame was not even close to Rivera's, but I thought she was more gifted than he was."

It was inevitable that Frida would be compared to her world-famous husband. They were a study in contrasts—a brilliantly mismatched pair, with their pet monkeys and dogs and birds: enormous Diego, with his hats and his feathers and the parrot on his shoulder or his head, and tiny Frida, in her long Indian skirts, and ribbons and flowers in her hair. For a while, their flamboyant life-style, the likes of which the quiet, restrained Frenchman had never seen, seemed to him like a true blending of art and life.

The quarrel between Trotsky and Rivera brought all social contact with the Riveras to an end. For a time, Comrade Van continued to act as the go-between, at first in futile attempts to restore the friendship and later, to settle what were designated as "business matters." In fact, these negotiations had less to do with business than with injured feelings and pride. The most immediate and pressing problem was the blue house, for, since they were no longer allies, Trotsky insisted that he could no longer accept Rivera's hospitality by living in his house. Van Heijenoort was, for once, equally adamant, insisting that they could not move until they found another house that was suitably secure. Trotsky then insisted that he had to pay rent until such a place was located. Insulted, Rivera at first refused the money, then he accepted it, then he refused it again; in the end, he accepted but gave the money to *Clave*, the review to which he and Trotsky had contributed (and argued about in the first place). It took van Heijenoort several months to locate another safe house and to see that the necessary remodeling was done to improve its security. Once that task was accomplished and the group was finally installed in a newly fortified residence on nearby Avenida Viena, not even Van saw Rivera again.

With Diego now on Trotsky's enemy list, Frida, too, became an untouchable. According to the principle of undivided loyalty, there was no way that van Heijenoort could continue to see her, even clandestinely, after that. As it happened, the immediate impact of his loss was diffused by her continued absence. In January, 1939, at André Breton's invitation, after several months in New York, Frida went directly to Paris for the opening of her first show in Europe. While she was there, both Trotsky and Rivera wrote to her of their conflict, each one giving his version of how the argument came to pass. The self-righteous Trotsky went so far as to ask for her help in making Diego see the light. Although he acknowledged that Diego's resignation would be damaging to the Fourth International, he made the amazing statement that it would also mean the "moral death" of Diego and that, without the Trotskyite affiliation, "I doubt whether he would be able to find a milieu of understanding and sympathy not only as an artist but as a revolutionary and as a person."

Frida, of course, had no worries or doubts about a milieu for Diego and, as Comrade Van predicted, she supported her husband as she always did when politics was at issue. She was far less conflicted than Van was in his support for Trotsky, since art, not politics, was her life. For van Heijenoort, personally, losing Frida was even worse than losing Diego as a friend and political ally. More than a helper, a lover, and a "confessor," she had been one of the few people with whom he could drop his guard, have a good laugh, and be silly without wondering whether it was compatible with Marxism. In general van Heijenoort had nothing but contempt for anyone who gave the slightest support or sympathy to Stalin, but in Frida's case, he excused her later infatuation with Stalin and the Communist party, saying that she was sick and under the influence of the drugs she was taking, and that, anyhow, politics meant nothing to her. In other words, she was not rational, not herself, and therefore not accountable.

The split with Rivera marked a change. Although on an external level, Van's reaction appeared to be a repetition of past history—the ever loyal Comrade Van making the painful decision to swallow his objections and abide with his leader—this time there was a profound erosion of the halo effect. It wasn't that van Heijenoort was entirely sympathetic to Rivera, but he understood his point of view and his feelings, and Trotsky's assumption that he could tell Rivera what he could or could not say or do seemed to him to be totally inappropriate behavior towards the man who had been their generous benefactor and guardian for two years. What van Heijenoort had already perceived as flaws in Trotsky's character, particularly his inability to yield or compromise, now disturbed him enough to affect his views on Trotsky's intellectual and political positions.

Years later, reflecting upon whether Trotsky's personal rigidity and arrogance had triggered his own critical stance, van Heijenoort said:

> I can't say that the one was the *cause* of the other or exactly what the interaction was. I know that even as early as 1937 I could read his articles critically and see that some of his arguments were weak. And there were other things too, the fact of my being older and more experienced. Trotsky's best writings were in 1932 and 1933 when the issues were fairly clear, but by 1939, things were much more difficult to explain. The world was a *mess*. How was anyone to understand such a world? So my reaction was a combination of all those things, my personal feelings, the objective side of things, and the intellectual side. It was very complex. It was not disaffection, but I simply penetrated more and understood more. I took a more analytical stance; I began looking at his writings in a different way and I found inconsistencies.

It was during this anxious period, in April 1939, that Loretta "Bunny" Guyer, a young, extremely attractive, and lively New Yorker, arrived on the scene. She came with her friend Lillian who had proposed the trip. Both of the women were Trotskyites, and a major reason for their journey to Mexico was the opportunity to meet Leon Trotsky. The young women had heard from their comrades that a visit with Natalia and Lev Davidovich would be possible if they went through the right channels. A tradition of such visits from "ordinary people" had been established during the days of the Dewey Commission hearings. Trotsky also gave more formal seminars once or twice a year to groups of Americans who came to Coyoacán expressly to hear him lecture and to ask questions. The Trotskyites viewed both types of gathering as a means of attracting followers and increasing the number of activists; and indeed, on occasion a new secretary or guard was recruited from among the visitors. Also there was always the possibility that some of the Americans might be rich and, therefore, make big and much-needed contributions to their cause.

As they had hoped, Bunny and Lillian were invited to Coyoacán for an audience with Lev Davidovich, and they were as well received as their friends before them had been. Rich they were not, but, to make up for it, they were young, enthusiastic, and profoundly impressed by their close-up view of the great revolutionary. As for Trotsky's handsome young assistant, who was always present during such meetings, Bunny thought he was a knockout: intense, striking, extremely attractive. "He looked like a movie star—like Jean Marais, the star of all the Cocteau movies," Bunny said later. "Jean van Heijenoort could have been his double." On this occasion, without question, the acolyte upstaged the priest.

Van knew Spanish, but he had not yet learned English; Bunny spoke no Spanish, and did not know French. German had been her language in school, but she had disliked it and had learned almost nothing. But it didn't seem to make the slightest difference that there were few words between them; the chemistry was all-consuming.

It was a ripe moment, more than a year since Gaby and Jean had separated and a few months after the rupture with Rivera. The Trotskyites had just moved from the blue house to Avenida Viena. Also in April, Frida had returned from Paris and New York after a six-month absence, but Comrade Van was in no position to see her without risking great disfavor.

*Jean Marais, the star of Jean Cocteau's films. "Van could have been his double."*

(This was a particularly sad period in Frida's life. At Diego's instigation, they were in the process of divorce. There was much speculation as to the cause of their difficulties: was it his many amorous affairs, or hers? Because van Heijenoort had been out of touch with her, he was not privy to Frida's thoughts and feelings. As it turned out, the Riveras divorced only to remarry a year and a half later.)

The day after Bunny and Lillian's visit to Coyoacán, the smitten van Heijenoort called on the young women at their hotel and offered to show them around the city. Bunny had told him she was leaving to tour other areas of Mexico and he was afraid he might not see her again. Recovering from his recent losses, Bunny's vibrance and charm captivated him; and Bunny, too, felt that something far more serious than a vacation romance was brewing.

The young women left Mexico City as planned. When they returned, Van and Bunny met again and the fire was re-ignited. Within a few days they declared their love, and by the time Bunny was due to go home, Van had asked her to come live with him in Mexico.

Uncertain of what she wanted to do, Bunny went back to New York with Lillian. She said nothing to her parents, but confided in her older sister, telling her everything and swearing her to secrecy. Still not completely sure, she nevertheless began to consolidate her finances. When she asked her sister to lend her some money, she realized she had made up her mind. Within a few weeks of her return home, Bunny was on her way back to Mexico. She had decided that her life was there, with Jean van Heijenoort.

Bunny and Van were married in July 1939, very soon after she returned to Mexico. (He was able to obtain a Mexican divorce without Gaby's presence.) As Gaby had done before her, Bunny joined the Trotsky household and became Natalia's domestic assistant in the kitchen and elsewhere. Van's second wife resembled his first in some physical features—she was small, dark, and lively—but there the similarity ended. Bunny was younger, quieter, less confident, and eager for instruction; she deferred to Natalia Sedova in all things. There was no argument or tension between them that she was aware of, and she simply tried to make herself useful in every way that she could. Looking back on that period of her life she said:

> I think innocence on my part kept me from doing anything that would cause trouble. I didn't have to think about being careful, I just was. In New York, before I went to Mexico, I had thought of myself as being grown up and politically aware. I thought I knew the score, but with Jean I felt naive, like a little girl, and very much in awe. It wasn't that he was older—it was only six years—it was the learning he had, and the living he had done.

Bunny was thrilled with her situation in Mexico. She was in love, she found her new home and its occupants exciting and fascinating, and she was enthusiastically ready to settle down and devote herself to the Trotskyite circle and its causes. In contrast to her life in New York as a part-time art student struggling through the Depression and supporting herself by working at odd jobs, she now had a strong purpose. And Mexico was working its charm on her as it did on all newly arrived visitors. Her husband, on the other hand, was in a quite different frame of mind. He had already had two and a half years of exotica and seven years with Trotsky. He was under what he described as "constant tension, twenty-four hours a day," and he was in desperate need of a breathing spell. "I needed to be on my own feet," he said, "and I needed a personal life of my own."

Consciously or not, by marrying Bunny, Van had provided himself with a legitimate means by which he could divorce himself from Trotsky, not ideologically or politically, for he was not yet thinking of that, but physically and geographically. At that point he knew, as never before, that he needed space and distance. As the husband of an American, he realized he had a way out, another place to go, and someone to go with. So, only a few short months after Bunny Guyer made her momentous decision to leave New York for a new life in Mexico, she discovered that the big step she had taken was on a circle that led her right back home.

Yet Bunny did not feel misused or deceived when Van proposed that they move to New York. On the contrary, she wondered how she could have been so fortunate as to have Jean Marais's double fall in love with her, and she was ready and willing to go anywhere with him.

In the closing pages of *With Trotsky in Exile*, van Heijenoort explains briefly why he left Trotsky and what he planned to do in New York:

> In October, my departure for the United States was decided upon. I had lived so many years in the shadow of Trotsky that I needed to be by myself for a while. I was just supposed to spend a few months in the United States. After that, we would see.

*Sieva (left), Loretta "Bunny" Guyer (center), and Natalia (right). 1939.*

The implication here is that his departure was something like a corporate decision, with everyone agreed that Comrade Van needed a vacation, and that most likely he would be coming back. Nowhere in the book is there a mention of Loretta "Bunny" Guyer or of van Heijenoort's marriage to her. The fact is, in his memoir van Heijenoort rarely focuses on the personal.

A second paragraph, four times as long as the first, deals with a barely possible political *raison d'être* for Comrade Van's move to New York. It is a serious, "all business" analysis:

> I left the house in Coyoacán early in the morning of November 5. On the evening before, I had my last conversation with

> Trotsky. We spoke about the situation in the American Trotskyite group, which was undergoing a crisis. The group was divided between a majority centered around Cannon, and a minority, led by Shachtman and Burnham. Trotsky feared that Cannon, with whom he was politically allied, would tend to replace the discussion of political differences with organizational measures, thereby precipitating the expulsion of the minority. "Cannon has to be held back on the organizational plane and pushed forward on the ideological plane," he told me. This was a little like the advice he had asked me to give to Raymond Molinier in Paris in August 1933. In this last conversation, Trotsky gave me no actual instructions for New York which my status as a newcomer would not have allowed me to apply. He simply explained the situation and in which direction I should move, according to my abilities.

Having a political mission in New York kept Van legitimate in his own as well as in Trotsky's eyes; the common purpose kept them together. In later years he would insist that "there was absolutely no *éloignment* [distancing] on the political plane," and as proof he offered the evidence that for five more years he continued to be a professional Trotskyite activist in New York. The benefit of being in the United States was that he would become more independent, more directly involved; the disadvantage was that he would have to work for his principles in almost total isolation.

Bunny, a native New Yorker who knew her city well, left Mexico first to find an apartment and a job. She did everything in her power to pave the way for her husband's arrival, but New York in November was far from the tropical paradise of Coyoacán. This "new planet" was cold, noisy, and hostile, and adjustment there was going to be much harder than Comrade Van had ever anticipated.

# III

Chapter Ten

# Out of the Shadow?

The United States: 1939–1945

For the first few years, I was totally lost in New York, crushed by the city, the language, the buildings. It was very strange, very disturbing. In Mexico, Coyoacán had been a rather country-like place, full of trees, and then, all of a sudden, I am living on Lexington Avenue and I have no contact at all with nature. It was Bunny who helped me, saved me. Just the fact of having her near me in that huge city.

JEAN VAN HEIJENOORT, *1984*

FOR ME, FALLING in love is easy. It's leaving that's hard." Jean van Heijenoort was speaking of Trotsky, when he used those words to describe how he felt about his separation from the Old Man. Later, he would say the same thing about his emotional entanglements with women. For seven years Comrade Van had lived, breathed, and taken sustenance with his leader, following his every thought and making allowances for every misstep. "After all that time," he said later, "I felt I had to get out from under the shadow of that man. I was twenty-seven years old. I needed to find out who I was."

Feeling the need for independence was one thing, achieving it quite another. New York in November 1939 was a cold, dark, dreary city filled with noisy people always in a hurry: an environment hostile to life, Van concluded. In his description of himself as "totally lost and crushed," the self-assurance and confidence the young revolutionary had acquired in Mexico seems to have vanished in the smoke of the metropolis.

Looking at him from the outside, however, one would hardly have noticed. Within a few months of his arrival in New York, the handsome Frenchman from Mexico had a part-time job teaching French at Berlitz and was engaged in full-time political work as Trotsky's ambassador, meeting with the American Trotskyites and corresponding with the Trotskyite groups in France and Europe. In the past seven years he had grown into the title of "Comrade Minister of Foreign Affairs" that Trotsky had bestowed upon him in Turkey when he was a green recruit of twenty.

Joel Carmichael, the editor of *Midstream* and a fringe member of the New York Trotskyite group in the early forties, remembered van Heijenoort's first appearances in New York very clearly:

> I used to see him at parties, in the early 1940s. He was very famous and glamorous in the New York political left because he had just come from Mexico—from Trotsky. People would whisper, "That's Trotsky's secretary." He was *very* reserved, wouldn't say much, and would just drift away from a group. Then there was this gossip about his affairs with women. So this handsome man who seemed absolutely glacial, turned out to be bubbling underneath like a volcano.
>
> I knew him for more than forty years, but to this day, I haven't the slightest idea what Van thought about Trotsky or Marxism or politics in general.

Van Heijenoort was not above responding to admiration when he was, as he put it, "on top of things." In Mexico he had felt important and perhaps even glamorous because of his association with the Riveras and their influential friends. He had had a small taste of power and found it agreeable. He had even been given a commission as commandant in the Mexican police force, which he thought was ironic and very funny—"I, who have passed through the hands of most of the police in Europe," he wrote to André Breton (who then turned around and used the phrase when he spoke of Comrade Van's bravery). But New York was totally different from Mexico. Van was miserable, although at first, only Bunny was aware of it.

The primary source of his dismay was his relationship with the leaders of the Socialist Workers party, the major Trotskyite group in New York. As Trotsky's emissary and now the newly appointed International Secretary of the Fourth International, Comrade Van had expected to be received with enthusiasm by these comrades. Instead he soon discovered that his supposed friends had views and priorities at variance with his. If they were not exactly hostile, they were indifferent. His concerns were global, whereas those of the Socialist Workers, in his opinion, were narrow and parochial. The Soviet–German non-aggression pact had been signed in August of 1939 and Germany had invaded Poland in September. He was deeply worried about the war in Europe, but his American comrades were focused on party politics in New York.

Nevertheless, he continued to work in liaison with the New York Trotskyites, and he wrote regularly for their journal, *The*

*Fourth International*, on the progress of the war and the various political ramifications of events in Europe and North Africa. Stubbornly, against the odds, he persisted in his attempts to keep what he conceived of as the true Trotskyist position alive. This was a problem he had to deal with in the framework of his own poverty. Since he had no financial support for his projects, he was in the demeaning position of having to swallow his anger and ask for aid from an unsympathetic boss. Describing his situation later, he said, "I would go to Cannon [James Cannon, then the leader of the SWP] and plead for money to buy the stamps and stationery necessary to maintain my contacts abroad. He would sit there, scowling and saying, 'mmmm, hmmm' but otherwise, not a *word*, as I explained how much I needed and why it was necessary." Grudgingly, Cannon doled out small amounts of cash, perhaps twenty dollars a month, but it was a humiliating experience.

There was, however, one great consolation that the inhospitable city offered to the sorely troubled van Heijenoort: the New York City Public Library, an island of quiet and culture, on Forty-second Street. A public library where anyone, rich or poor, educated or not, could enter to read or write or dream was a completely new phenomenon to him. There was no such place in either France or Mexico. When to his amazement and joy, Bunny told him about the resources at Forty-second Street and initiated him into its wonders, his life was changed and he almost forgave New York its sins. When he said, "Bunny saved me," this was one of the things he was talking about.

"I would go there every day that I could," he said later. "I'd be there when they opened at nine in the morning and I'd stay there all day, until five or six in the evening, absorbing, just absorbing. It was my greatest discovery." More than a discovery, it was his salvation.

It was as if Mexico had been the desert, not the tropics, and he had to make up for years of drought by soaking up the plentiful waters of knowledge and information now suddenly available to him. And Forty-second Street was only the beginning. Now that he knew such wonders existed, when his Berlitz job took him to Baltimore and Philadelphia, he sought out the university and public libraries in those cities, too. Later, in Boston and Cambridge, or wherever he happened to be, he would make his home and his office in the library and spend as much time there as he possibly could. Repeating the experience of his youth, he again found comfort and security in books, in intellectual work and its trappings, and above all, in the certainty of mathematics.

> At Forty-second Street, I had my seat; I would always sit in the same place. I was absorbing all kinds of things that I had no chance at before—such as Whitehead and Russell's *Principia Mathematica*. I mean I didn't *read* it, I wasn't ready for that yet, but I looked into it. Gide on Dostoevsky, the Russian criticism of Einstein—that kept me occupied for a few weeks. And then I was reading a lot of political information on the Spanish Civil War, and on the war in general. I was not reading novels, except maybe once every six months.

Professor Sidney Hook, then at New York University, had a story somewhat at odds with van Heijenoort's assertion that he had not had access to reading materials. According to Hook, the Trotskyites in Mexico used to send lists of books, and other wished-for items, to their supporters in New York, who would then do what they could to fill the requests. At one point it was noted that Trotsky was asking for some very serious scholarly stuff: *Principia Mathematica,* to be precise. Then the word went around that the Old Man had wider interests and greater brilliance than even his most passionate admirers had imagined. No one suspected that the secretary who wrote the letter might have had something to do with the nature of the request until many years later when the secretary turned up as a student at New York University. (Of course it is possible that the requested *Principia* never arrived or that Comrade Van never had time to look into it in Mexico, so that his first *real* encounter with Russell and Whitehead may indeed have been in New York.)

Besides absorbing, van Heijenoort was also producing. At his seat in the library, the quiet man led more than a double life. While one part of him was reentering the world of abstract thought, the political side was divided into a half-dozen or so different authors, all of whom were busy producing copy for the *Fourth International*, the New York Trotskyite journal. Using the names Vladimir Ivlev, Karl Mayer, Marc Loris, Daniel Logan, Jean Rebel, and Ann Vincent (the last, when he began to get draft notices for his masculine pseudonyms), he wrote a number of articles which were, for the most part, descriptions and analyses of what was happening abroad. *Europe Under the Iron Heel*, written by "Marc Loris," is one example, an essay which sets out "to provide information for the non-European reader on the situation now existing on the continent which for centuries was the guide of mankind." In *The Political Misadventures of the French Bourgeoisie*, also by "Loris," he discusses the Vichy government, the resistance movement, and the emergence of Charles de Gaulle. His major focus was Europe, but he also wrote about Africa and Lebanon. Between 1941 and 1944, van Heijenoort wrote more than twenty articles based upon the research he was able to do in the library as well as information that he received from informants abroad.

As Secretary of the Fourth International, Comrade Van also carried on a large correspondence, gathering and exchanging information with Trotskyites in other countries of the world. Since almost all of Europe and by the end of 1941 the United States, too, was at war, maintaining contacts and transmitting messages was a much more complicated affair than a drop in a mailbox. Holding on to his old habits of secrecy forty years after the fact, van Heijenoort was initially close-mouthed about how he had managed to communicate abroad. He would say no more than, "Oh you know, there were sailors." Only after insistent questioning did he finally explain that the Socialist Workers party had a network of seamen, rank-and-file members of the SWP who had joined the merchant marine instead of waiting to be drafted into the army or navy. These men sailed all over the world and acted as couriers. They were able to deliver and receive letters, newspapers, journals and other documents as well as messages which gave news about what was happening in the areas they reached. And occasionally even more direct sources of information were tapped. One sailor, on the Murmansk run, surely deserved the Man-of-the-Year award from the Socialist Workers party because, according to van Heijenoort, he managed to become the lover of a Red Army General, a woman, who, in intimate moments, unwittingly passed on all kinds of insider information.

Finally, the Forty-second Street Library served as van Heijenoort's private club, his home away from home, his meeting place for social and intellectual discourse with the other members of the Forty-second Street crowd of readers, writers, thinkers, and hangers-on. Among these were Nicolas Calas, a surrealist art critic who would later become a professor of art history at Fairleigh Dickinson University, and Lionel Abel, who was translating Rimbaud and Sartre, writing plays, and would become an English professor at the State University of New York. Both these men spoke French, an attribute of great importance to the exiled van Heijenoort, who felt the assurance of being understood only in his own language. Calas, of Greek heritage, had lived in Paris from 1930 to 1940 and was an even more recent immigrant than van Heijenoort, and Abel, too, was conversant with French literature and culture.

Nicolas and Van met almost daily and became close friends. "We used to spend endless hours, the two of us, walking and talking in the corridors of the Forty-second Street library," Van said. (Even years later, his admiration for the library was reflected in the reverent way he sounded the words, "Forty-second Street Library." )

> We talked about *everything*, about politics and political situations, about Trotsky, about New York, about surrealism, about the Marquis de Sade, about Breton and Rivera and Frida, about Don Juan, Casanova ... just anything.

At first, their conversations about Frida Kahlo and André Breton were a felicitous point of connection. Both Calas and Abel had known Breton in Paris (and they were all soon to see him again in New York), and, of course, van Heijenoort had known Frida very well in Mexico. But ultimately, van Heijenoort was irritated by Nicolas' view of Frida and her art, and it became a point of contention. Calas had encountered

Frida late in 1938, at the opening of her sensational show at the Julian Levy Gallery in New York. Everyone who was anyone in the art world had come to that show, and Frida, at her flamboyant best, had impressed Calas and others with what he called her "theatrical quality and high eccentricity. She fit completely the surrealist ideal of a woman," he said.

Van Heijenoort's response to this assessment was the same "that's crap" that he had already rendered as judgment upon André Breton's attempt to classify Frida. He suspected that Calas was simply repeating, if not plagiarizing Breton. "He [Calas] just wanted to capture her for the surrealists, but you couldn't do that with Frida. You couldn't fit her into your idea of something. She was independent of all that."

Their arguments about Frida and art were only the beginning of their differences. Eventually Van and Calas argued on political matters as well, with such violence that their friendship could not withstand the strain. "Finally Nicolas just went too far," Van said, "I had to break with him." On the other hand, Van's friendship with Lionel Abel, which was less charged, continued into the period when they were both settled into academic positions.

In the everyday world, Van had his wife, Bunny, to "save him" and to deal with life in the big city. He felt he owed his very survival to her, but their relationship was not on what he called "the intellectual plane." She was warm and loving, provided home and hearth, and, not incidentally, was the major breadwinner, but she did not speak his language and, rightly or wrongly, deep down he felt that she really didn't understand what he was all about. Whether he knew what *she* was all about is another matter. Understanding the feelings of even those people who were closest to him was never his best talent, nor was it his highest priority.

To say that van Heijenoort felt physically and emotionally adrift in his first years in the United States is understating the case, even though it appeared that his life was organized around a routine of teaching, library work, and political activity. The physical drift was concrete: he and Bunny were constantly on the move. They changed apartments because of noise or inconvenience and changed cities for economic reasons. But, besides the fact that none of the apartments they lived in ever satisfied Van, he also thought it was a good idea not to stay in one place long enough to become known and identified. They moved between New York, Philadelphia, and Baltimore depending upon where the Berlitz job opportunities were best. As an experienced hair stylist, Bunny had no problem finding a job wherever they went. Baltimore and Philadelphia were cities where van Heijenoort found some measure of tranquility, where he could breathe more easily and feel released from the grinding tension of New York. The architecture in both cities pleased him; it was on a human scale. As always, he had his libraries and happily found himself a place in them. He particularly liked the public library in Philadelphia, a graceful red brick building with white trim, built in 1892.

The van Heijenoorts happened to be in peaceful Baltimore on August 21, 1940, when they learned the shocking news of Leon Trotsky's assassination. Van, on his usual morning walk, spotted the black headlines on the front page of the *New York Times*: "Trotsky Wounded By Friend In Home—Is Believed Dying." He returned home immediately, and he and Bunny sat listening and waiting at the radio. Later in the day, they heard the dreaded final words: "Leon Trotsky died today in Mexico City."

"Darkness set in," is the last sentence in van Heijenoort's memoir of his years with Trotsky. Those dramatic words reverberating against his usual terse style of factual reporting or setting the record straight, make clear his intention to underline and heighten the effect of the tragedy. "Darkness" for Trotsky, but also darkness for the apostle who had loved Lev Davidovich so deeply. It was truly a black time for him and for Bunny who suffered with him. For days they just sat in a park, stunned. And since silence was Van's customary reaction to sorrow and anger, they barely spoke.

Although Van had had legitimate reasons for leaving Trotsky and Mexico, when he learned the grisly details of Trotsky's assassination, he could not keep from blaming himself. Nothing shook his certainty that had *he* been on the scene, he would have been able to prevent this particular attack. He had chosen to protect Trotsky, and for seven years he had been successful in that endeavor. But after his departure, in nine months, there had been two attacks on Trotsky's life; the first had failed but now, the always anticipated worst had happened. The fact that Van's once simple faith in his hero had diminished, and that his state of mind regarding the living Trotsky had been, of late, a tangle of personal embarrassments, unexpressed disagreements, and bottled-up resentments, partly submerged and unacknowledged, made his guilt and devastation all the greater.

Trotsky's murderer, an agent of the GPU, was Ramon Mercader, a Spaniard who posed as a Belgian Trotskyite under the aliases Jacques Mornard and Frank Jacson. It was precisely this disguise, van Heijenoort said, that should have marked him as an imposter. "I would have recognized him immediately because his French would have had an unmistakable Hispanic rather than a Belgian accent. That would have been enough to make me suspicious. I would *never* have let Mercader into the house to talk to Trotsky alone." Van Heijenoort realized that once Stalin had decided that he wanted to be rid of Trotsky, nothing and no one could have stopped the attempts at assassination. But Mercader alone, he always insisted, "I'm sure I could have prevented."

Mercader-Jacson had plotted his attack over a period of years. In Paris early in 1939 he had introduced himself to Frida at the exhibition of her paintings which had followed her New York show. Telling her he was going to Mexico, he asked if she could arrange for him to meet Trotsky. She refused, but only because by then the Rivera–Trotsky rupture had occurred and the great friendship was no more. But Mercader was more successful in his other moves. In a plot which sounds like part of a spy novel, he became the lover of Sylvia Ageloff, a young American Trotskyite visiting in Paris, and used her to gain access to Leon Trotsky. (According to one account, Sylvia was given a free trip to Paris and set up for Ramon to seduce her.)

By 1940 both Sylvia Ageloff and Ramon Mercader, calling himself Frank Jacson, were living together in Mexico. Sylvia's sister, Ruth, also in the movement, had volunteered her services to help Trotsky during the Dewey Commission hearings of 1937, and now Sylvia, too, offered to work for the cause. She was immediately accepted and became a regular in the house on Avenue Viena. As Sylvia's "young man," Mercader soon met

Trotsky and assumed the role of friend and sympathizer of the movement. He pretended to be writing an article on the Soviet Union and solicited Trotsky's advice. Thus he was admitted, alone, to the Old Man's study. As Trotsky began to read the manuscript, Mercader attacked him with an ice axe he had hidden in his coat, plunging the weapon into his skull. The powerful Lev Davidovich had the strength and fury to fight back, to bite Mercader's hand, grab the axe, and to cry out. Natalia and several others rushed into the room; the murderer was caught, but his mission was accomplished. The following day, in a hospital in Mexico City, Leon Trotsky died.

To his bosom comrades, Trotsky had predicted that, in the event of war, Stalin would make an attempt on his life, and, also, when it would be made. The Old Man even had a scenario in mind for how the deed would be done: he had imagined a demonstration staged outside the villa, during the course of which some of Stalin's professional agents would overwhelm the guards, enter the house, kill him and burn his papers. He was right about the time but quite wrong as to method. The job was done quite simply, by one man feigning friendship. When it came to betrayal of that kind, Trotsky's pride in his personal and intellectual magnetism limited his imagination.

In May, 1940, only three months before Trotsky's murder, there had indeed been an unsuccessful assassination attempt that was closer to the Old Man's nightmarish fantasy. This sortie—a wild pre-dawn raid by a gang of Mexican Stalinists whose leader was the painter David Siqueiros—had been bungled by the assassins rather than averted by the bodyguards. The attackers took one guard as hostage, tied up the others, and entered the house with pistols and machine guns. Trotsky and Natalia awakened to what they thought was the sound of holiday fireworks, but quickly realized they were hearing machine-gun fire. In the dark, they crept out of their bed and flattened themselves against the wall, next to a corner. Although bullets riddled the bedroom, the doors and windows, and even the very bed the Trotskys had just vacated, to everyone's utter amazement they emerged alive and unscathed.

Trotsky's explanation for their miraculous escape was that the assassins, hearing no sounds, and having shot what they thought was the *coup de grace* into the shapes of bodies created by the rumpled covers on the bed, assumed they had hit their mark. Other political analysts have claimed that the intent was only to terrorize and not to kill, that Stalin wanted the actual assassination to occur at a later date; and still others insist that the whole episode was staged by the Trotskyites themselves. Trotsky's grandson Sieva, who was in the adjacent bedroom and was shot in the foot, discounts both these views as the Communists' alibi to avoid prosecution for attempted murder. Even Diego Rivera was under suspicion for a short while, because his break with Trotsky had been widely publicized and was considered as a possible motive. He went into hiding for several weeks but soon it was evident that he was not involved. In the end, Siqueiros and his followers were arrested, tried, found guilty, and sentenced to jail.

Upon hearing news of the aborted assassination, Comrade Van had proposed to return to Mexico to take charge of security as he had done so successfully in the past; but Lev Davidovich refused his offer. In a touching note, the Old Man wrote, "It would be really too cruel to force you to return to this prison. Now, after the reconstruction, it has become a genuine prison, not in the modern manner, it is true, but rather more like those in medieval days." So Van remained in the United States and the inevitable happened. Trotsky's rueful comment makes it clear that he felt himself and those with him to be the prisoners of their own need for protection, but at that point he could not have known that the "medieval" walls were not enough to keep out even one resourceful intruder.

* * *

A quarter of a century had not erased van Heijenoort's memory of his own young father, coughing up his life's blood. Time would never do that. If anything, the emotions surrounding his father's end, to which he bitterly referred as "my arrival in this world," had been elaborated upon and enlarged, as the subsequent events of his life seemed to him to have been determined by that tragedy. Although he was not there to see it, he had no trouble conjuring up the awful scene of Lev Davidovich's murder. His spiritual father, the vigorous hero he had admired more than anyone else in the world, the Old Man who had shaped his life as "Comrade Van" and named him "Comrade Already," had died, also in blood. This was what all Van's effort of the past eight years had come to.

As soon as they could, the van Heijenoorts left Baltimore and returned to New York. By the fall of 1940 Van had renewed his political activity there as the chief, if not the only, representative of Trotsky's brand of Trotskyism in the United States. He took this responsibility more seriously than ever, but at the same time, for reasons beyond the obvious fact of Trotsky's death, he was deeply depressed. He no longer held any illusions about the possibility of having an effect on the political situation in the United States, or even within the Trotskyite inner circles with which he was supposedly allied. World War II was raging in Europe; Hitler and Mussolini were swallowing one country after another; and finally, on the personal level, because of the war, he felt more cut off than ever from his family and friends in France.

There are two tragedies in this world, said Oscar Wilde and Bernard Shaw, in slightly different ways: one is not getting what one wants, and the other is getting it. Van Heijenoort had wanted to be independent, to free himself from Trotsky, to be someone in his own right, not another man's shadow. Now, most unhappily, those desires were satisfied, but what did he really have? First, on the most superficial level, a job with Berlitz, teaching the most simple-minded French language courses. Next, on a level he had thought would be important, a position: International Secretary of the Fourth International. But the sad truth was that the grand title was almost meaningless, and nothing could gloss over the fact that the constituency he represented in the United States was pitifully small, perhaps not more than a few dozen people. Van Heijenoort's wish "to be by myself for a while" had been fulfilled with an isolation deeper than he desired.

However vague the plans had been, when Van left Mexico, he had thought his separation from the comrades in Coyoacán was temporary. Now, with Trotsky gone, there would be no going back. Almost certainly, if France had not been occupied by Germany, van Heijenoort would have returned to France, with or without Bunny. Even with the war, he thought seriously of joining General de Gaulle's resistance movement or of going to Sweden to participate in the European conflict from there. In the end, he did not see a way to get to Europe and be useful, so he stayed where he was. But he was a man twice cut loose from his moorings, still desolated and guilt-ridden over Trotsky's death and unable to accomplish what he saw as his task in this foreign country.

* * *

After Gabrielle Brausch van Heijenoort left Mexico at the end of 1937, following the infamous episode with Natalia Sedova in the Coyoacán kitchen, she had returned to her job and her political activity in Paris. Jeannot had been settled with his paternal grandmother Hélène in Creil, where he basked in the attention and affection his absent father had once yearned for. Gaby visited on weekends, saw that her boy was thriving, and concluded that all was as well as could be expected.

The onset of World War II changed everything. As soon as Germany invaded France and the Vichy government was established, Gaby and Maurice, her new companion, took Jeannot and fled to the south, first to central France and subsequently to Marseilles, where they became active members of the Free French Resistance movement.

It was there, in Marseilles, that van Heijenoort, using his connections with the cadre of Trotskyite merchant seamen, succeeded in locating Gaby and communicating with her to discuss the safety of their child. The result was that with the help of the International Red Cross, Gaby and van Heijenoort were able to arrange for Jeannot's passage to the United States. Van bought the ticket, got the visa and even offered his ex-wife his help by proposing to pay for her passage as well, if she would come. She declined, saying that her place now was with her comrade Maurice and with the Resistance. In May 1941, Gaby and Jeannot left for Lisbon, and after a tense and uncertain journey and a heart-rending farewell, Gabrielle Brausch put her son into the hands of the Red Cross. Then she returned to what she saw as her duty in France and, as it would turn out, to disaster.

Meanwhile the intrepid six-year-old Jeannot made his second Atlantic crossing, without family, but in the company of hundreds of other refugees aboard the ship. Forty years later, he remembered the trip on the boat as fun. "What does a little kid know about war and danger," he said. The truth is, with his background, he knew far too much, but it didn't seem to faze him as a child, and as an adult, he maintained a rather cheerful, unsentimental perspective on his situation:

> I remember the declaration of war in 1939. I was in Nogent [just outside the city limits of Creil] with my grandmother. There was no more milk for breakfast. In Marseilles, there was nothing to eat. Boats couldn't go out to fish because of the war and there was no food coming into the city from the country. Then, my mother took me to Lisbon and sent me to my father in America.

Marvelous what a boy will think of as "fun": separated from his mother and Maurice, and from his grandmother, the person he loved the most; traveling to a strange land, across the wide ocean, on what might have been the last refugee boat from Portugal; going to a father who was no more than a vague memory and a stepmother he did not know—all of that was held in abeyance, and memory glossed the ocean voyage as a good time.

The stranger–father who came to meet Jeannot at the dock in New York and then took him home was "kind and comforting but not very talkative." And except for the first day, what talking he did was in English.

> My father spoke no French at all to me in the United States, and so I had to begin to speak English right away. He thought that was better because in 1941, in New York, all foreigners were suspect, and also because he knew I was going to be spending more time with Bunny than with him. It was Bunny I had to speak to, and she didn't know much French.
>
> I had no idea what my father was doing in New York besides teaching French. He never told me anything about his politics. All that was a fog. It was my mother [Gaby] who told me

> everything when I went back to France. In America, as I saw it, I was supposed to become just an ordinary kid, reading Superman comics, eating popcorn and drinking Coke.

His efforts at becoming that "ordinary American kid" were a ringing success. He went to school, learned English rapidly, and forgot how to speak French. Later, at thirteen, when he was told by his father that he had to return to his mother in France, he felt as if he were being sent to a foreign country where people spoke a language he did not understand, and he did not wish to go. Seven years in the United States had Americanized him more than forty-seven years would ever do for his father.

Nevertheless, Jeannot was understandably confused about how to answer the inevitable, self-defining, "What are *you*?" question that children ask each other. He really didn't know. In a very short time he had come to think of Bunny as his mother, and in fact she assumed the role of mother *and* father, because if van Heijenoort wasn't working, he was in the Forty-second Street or some other library, writing and "absorbing." Bunny's relatives, who were, as she put it, "more or less Jewish," became Jeannot's family, and "mother" and son made frequent visits to the family in the Bronx. Still, Jeannot found it hard to identify himself as New York Jewish since his French father with his Dutch name was decidedly detached and totally different from that group.

Rarely did van Heijenoort visit his wife's family. When he did, he was taciturn and removed; it was almost worse than his not being there. His in-laws were not the only ones to bear the brunt of his chilling silences. As Jeannot later recalled,

> Sometimes, for days he wouldn't talk at all, not to anyone, or he would only talk about what he thought were important things, like mathematics or politics. He would never just chat around to anybody about anything. My father never really took much time with me. He never helped me with my homework and he never told me about himself.

Not many people, and least of all his young son, would understand this denial as a reflection of Van's own misery.

There were, nevertheless, a few things that Jean, *père*, did do with Jean, *fils*. He taught him how to use a typewriter, and he built him a telegraph machine and taught him Morse code. Together, father and son built some little radios. Always gifted when it came to technical matters, the ex-Comrade Technocrat passed his talent on to his son. In later years, the younger van Heijenoort would say it was one of the things he appreciated most about his father.

But aside from his "good hands," it seemed to Jeannot that his father knew nothing at all about normal life and everyday situations. Even as a child Jeannot understood that his father had cut himself off from mundane affairs, either because he didn't know how to cope or because he didn't want to know. Nor did he know how to treat a six-year-old or a twelve-year-old or how to have a plain ordinary conversation with anyone. As a boy, Jeannot could not explain why it was so; all he could do was learn to live with it.

It seems strange, and even cruel, that van Heijenoort brought his son to New York and then proceeded to ignore him. He had gone to great lengths to provide refuge, but once Jeannot was with him, van Heijenoort essentially turned him over to Bunny, as if it were not his job to deal with a child. The question is, how could he, a "good" man, who thought he was devoting his life to making a better world, account for such behavior?

Van Heijenoort himself was not disposed to examine the origins or motivations for his relations with family in the same way he thought and commented upon the reasons for the political or scientific choices he made. Still, certain facts do bear upon his relationship with his son, the most obvious of which was his simple ignorance about what fathers did. Growing up in his "milieu of women," as he called it, had offered no teaching in fatherhood. He had neither experience to call upon nor a model to emulate. Perhaps in some situations negative experience is a motivating force, and a person resolves not to replay history but to do better unto others than he was done by; but not in Jean van Heijenoort's case. He wanted his boy to be safe and well cared for, but he left it to Bunny, a woman, to do the actual caring, while he persisted in and refused to be distracted from the manly political business of changing the world. (Models were not lacking for this way of dividing responsibility.)

Had his political work been in any way successful, perhaps his home life might have been happier. Instead the picture was as gloomy as it could possibly be, and he became ever more depressed as he contemplated his pitifully ineffective efforts. At every point in its history, the Socialist Workers party had been one of the smallest minority parties on the political left. Its voice was sometimes strong because many of its members or sympathizers were articulate, often eloquent intellectuals, but its impact was negligible. As in the *Ligue Communiste,* the Trotskyite organization in Paris, internal relations within the SWP in New York were always tense. There was endless disagreement and infighting between personalities and positions, between the Cannonites and the Cochranites and the Shachtmanites and Goldmanites, and, as in Paris, these fights often ended in splits and the creation of even smaller groups. The situation was made infinitely worse when the United States and the Soviet Union became uneasy allies during World War II, and the United States government did not want to do anything to antagonize Stalin unnecessarily. "Then we really were *les chiens lepreux* [the leprous dogs]," van Heijenoort explained, "outcasts hated more than ever by both the left and the right." Certainly some of his silence and coldness towards his son and everyone else can be attributed to his own despondency and emotional exhaustion during this most difficult period.

* * *

In the academic year 1943–1944, van Heijenoort took the first steps toward a return to the academic world and a career in teaching and scholarship at a level more advanced than Berlitz. He applied for and got two part-time positions: one in the French department at Bard College, which was then affiliated with Columbia University, and the other at Harvard University, as a teaching fellow in electronics in the pre-radar Officers' Training Program in the Graduate School of Engineering. He,

Bunny, and Jeannot moved to Annandale-on-Hudson, found a house on the Bard campus, and from there Van commuted to his second job in Cambridge to teach his classes at Harvard. Of course, teaching French at Bard posed no problem for him, and the content of the Harvard courses was basic electronics: "unclassified, very straightforward, and very easy to teach," he said later. He claimed he had learned most of the material for the course—which he called trivial stuff—by building those little radios with Jeannot, but, in fact, he had also taken some night courses in New York to learn everything he needed to know to qualify for the Harvard position. Furthermore, he had a life-long interest in electronics; then and later he always kept abreast of the most recent developments.

Although van Heijenoort would soon realize that both the United States immigration authorities and the Federal Bureau of Investigation had launched a full-scale investigation of his personal and political background, at that point he was blissfully unaware of the extent of their surveillance and the troubles that would follow. The ambience of Annandale and Cambridge suited van Heijenoort much more than New York's cold hustle and bustle. There he felt the strain of big city life dissipate and fall away. And Jeannot, too, later remembered that year

> ... as perhaps the happiest one of my childhood. In Annandale we had a wonderful, small old house between the baseball field and the woods. After Easter we moved to Cambridge, where we stayed in a huge old house on Kirkland street until September of 1944. I did quite a few things with my father. We had a vegetable garden which was a great success.

Living in New England was agreeable except for the weather. Van Heijenoort, with his predisposition to respiratory ailments, was subject to frequent and severe sinus attacks, which both he and his doctor concluded were brought on or aggravated by the cold and damp winter. Partly for this reason, the van Heijenoorts decided to move back to New York, but first, in Boston, Van had a sinus operation to relieve the general problem, and afterward his doctor suggested a few weeks in a warm dry climate as a prophylactic measure to speed his recovery. With no hesitation,

Van took his doctor's advice. Van was a brave man, always willing to take risks in regard to his personal safety, but that did not mean he was careless about his health. Quite the contrary: all his life he was a careful monitor of all his systemic functions. In this case, he was tending his respiratory system. So while Bunny and Jeannot returned to New York to search for yet another place to live, Van took the Greyhound bus to Arizona.

Armed with a chapter of Trotsky's book on Stalin which he planned to translate as he recuperated, van Heijenoort arrived at his destination. "Tucson at that time was small and very lovely," he reported years later, but his recuperation there turned into an unexpected misadventure:

> I rented a room in a private house for five dollars a week and prepared to settle in for a while. I did not know a soul in town, but I didn't care. I was there for the warm, dry weather, and I could work on my translation.
>
> One evening, I went out to buy some grapefruit, because they were so big and sweet and so cheap and so good for my health. As I was walking back to my room, two *huge* military policemen came from behind and grabbed me under the arms and said, "You are an escaped German prisoner; you are under arrest," and they took me to the police station.

Unknown to van Heijenoort, there was a large prisoner-of-war camp nearby from which, indeed, a dozen or more Germans had recently escaped. At the police station, Van protested that he was French and showed his alien registration card, but he was not believed, and, with the name van Heijenoort, it is not surprising. He did not tell the police where he was staying because he feared that if they went to his room and found the Trotsky chapter, which he had left lying on his work table, he would be in even more serious trouble.

> They put me in the county jail; it was so filthy—you wouldn't believe how filthy. And the food! In the evening, they threw an unopened package of bread through the bars. That was dinner; and in the morning, they brought a big pan of so-called soup, so rotten you couldn't even smell it, let alone eat it. That was breakfast. I had been in jails in Europe but nothing there was

> ever that bad. But bad food wasn't my only problem. All the other prisoners in the jail were convinced I was a German spy. They were rough types and ready to beat me up for that until I finally made them believe it was all a mistake.
>
> It took almost a week before I was called in and given back everything that had been taken away from me and told that I could go. I guess they had called Washington and found out I was telling the truth or else they found the real prisoners. A few days later, I met the chief of police on the street and he tipped his hat and said hello to me.

This friendly, "let bygones be bygones" gesture gave van Heijenoort a feeling of homeyness and safety, and thinking that the climate and the grapefruits were doing wonders for his health, he considered staying on a bit longer. He actually went out looking for a job. A "Dishwasher Wanted" sign in a window attracted his attention and he decided to apply. The boss was ready to hire him, but at the last minute he asked van Heijenoort where he had worked before. Van, in his naiveté, told him: "I have been teaching radar at Harvard [pronouncing it 'arvard]. Then that guy looked at me like I was crazy and he said, 'Get out!'"

The Greyhound bus took van Heijenoort back to New York. Bunny had rented an apartment in Greenwich Village, and Van was very pleased. The Village, with its cafes and bookstores, reminded him of Paris. It was the one place in the city that gave him a feeling of belonging. While he was still planning to spend some time and energy writing articles for the *Fourth International,* in the spring he found a job teaching at Wagner College, a Lutheran institution on Staten Island, where he was paid five hundred dollars for teaching two courses in elementary mathematics. Because so many able-bodied men were in the armed services, anyone with the skill to teach technical subjects was very much in demand and from van Heijenoort's point of view, teaching elementary mathematics was preferable to teaching elementary French at Berlitz. He had had enough of *le crayon, le stylo, et la plume de ma tante.*

By the rear door, then, van Heijenoort found his way back into the academic world he had abandoned. It was not quite the real thing as he had known it: at Wagner there was nothing like the rigor of the Lycée Saint-Louis, and at Harvard he had taught radar, not mathematics. Still it was, at last, another beginning and enough to whet his appetite for a bigger bite of academia. And so it was that in the summer of 1945, after passing an entrance examination which included an English exam, he enrolled as a graduate student in mathematics at New York University.

Chapter Eleven

# Another Life

## New York: 1945–1950

By 1948, the Marxist-Leninist ideas about the role of the proletariat and its political capacity seemed more and more to me to disagree with reality ... I pondered my doubts, and for several years the study of mathematics was all that allowed me to preserve my inner equilibrium. Bolshevik ideology was, for me, in ruins. I had to build another life.

JEAN VAN HEIJENOORT, *With Trotsky in Exile*

TO MAKE YOUR mark in the world of mathematics, so the myth goes, you must do a major piece of work by the time you are thirty years old. Jean van Heijenoort, who had just turned thirty-three, may not have heard the story *before* he enrolled as a graduate student at New York University, but he would become aware of it in short order. And, if his age alone was not worry enough, there were the thirteen years that had lapsed since he quit mathematics, "without a second thought," and plunged into politics. The return trip from the gritty air of everyday activism to the rarified atmosphere of the abstract would not be so swift or easy. It was to be a journey undertaken in stages, with more than enough time for reflection and doubt.

But whether or not he worried over the validity of the myth or about being "rusty," the solace provided by the academic world was real. At the very least, it was a world apart, a place where he could tune out his other problems and lose himself in the mental and physical details of being a student. Soon enough, it was an environment in which, once again, he found himself completely at home. His old habits of long and deep concentration stood him in good stead as did the broad and rigorous preparation he had received at the Lycée Saint-Louis in Paris. Immersed in books and ideas, he worked at catching up and moving forward. If he resumed his career as a student with the goal of becoming a professor because he had to do *something*, it was also true that, politics aside, mathematics was what he knew best and loved most.

Van Heijenoort's debut as a graduate student coincided with the end of World War II, a time of great vitality and ferment in the universities. The general mood of the country was one of elation, release, and renewal. People were putting their lives back on track, taking up where they had left off or starting anew with fresh resolve. In the world of mathematics, new fields of research were being developed, and new ideas and new theorems came pouring out. Stimulated and excited by all this activity, Van was immediately engaged, and before long he began working on problems in differential geometry. In the course of his studies, he was accepted as a doctoral student by Professor James Stoker, one of the most distinguished mathematicians at New York University.

During that period at NYU, there were a remarkable number of unusually gifted graduate students, such as Joseph Keller, Peter Lax, and Louis Nirenberg, all of whom would become world-class leaders in mathematics and its applications. In that elevated company, van Heijenoort was not a star and he knew it. It was also obvious that Professor Stoker, known to be a difficult task-master, did not look upon him as one of the brilliant young innovators of the late forties. But modest though he was, Van still had the necessary and crucial confidence in his ability to solve problems and to develop ideas. At the end of four years he wrote a very respectable thesis, *On Locally Convex Surfaces*, and in 1949 he was awarded the Ph.D. degree.

Including his time as a graduate student, and later as professor, van Heijenoort would remain at NYU for twenty years—until 1965. Many of his fellow students, who were also appointed to professorships, stayed on as long or longer than Van did. Yet these longtime colleagues knew almost nothing personal about him. No one knew where Van lived or with whom. No one knew that he had a wife, and a son from an earlier marriage living with him. When his daughter was born in 1950, neither cigars nor candy were passed, and only his closest friends were aware of that event.

According to Professor Joseph Keller, whose office was right next door to Van's,

> Van was a loner, an oddball; we never knew exactly how he fit in. During that period there was a whole group of us who had been students together: Peter Lax, Louis Nirenberg, and Harold Grad and Eleazar Bromberg and Eugene Isaacson and Cathleen Morawetz and quite a few others. There was a lot of mathematical interaction between us, and we would socialize a good deal, but

he was never part of it. So far as I can remember, Van never went out to lunch with us, and he didn't participate in the seminars very much. He mainly did his own thing. He'd go into his office and just sit there and work.

*With one exception*: there were these young girls, students from his classes, who often came to visit him in his office, and then, he would leave with them. I was very favorably impressed by that!

Another member of the NYU community, Professor Peter Lax, saw the "oddball" from a different angle:

> He seemed to have come out of left field, and he had this very interesting background, although I didn't know about it for quite a while. He was older than the rest of us, but that seems an odd thing to say, too, because he seemed ageless to me. He was reserved, very polite, and very European.

Lax, who had come from Hungary when he was eighteen, was striving to become "All-American" and to get rid of the traces of his own Hungarian accent. So he was particularly struck by van Heijenoort's strong French accent and the fact that *he* was doing the opposite: "I always had the impression that he actually *worked* to preserve it." There were some who agreed with Lax and thought Van cultivated his "Frenchness"—that it was an affectation or an unnecessary clinging to past mannerisms. But others disagreed. Van Heijenoort's son, for example, has said that his father made efforts to improve his accent. "After he died, my sister and I found books on English pronunciation in his belongings. It is not easy for a Frenchman to have a proper accent in English. My father was over twenty-seven when he started learning the language and he did not have a musical ear." But whether Van's Frenchness was cultivated or unshakable, it was something that Lax thought set his colleague apart and made him inaccessible. Having first placed van Heijenoort in left field, he ended by putting him considerably farther out, saying, "To me, he was like a man dropped from the moon."

Other colleagues had a more sympathetic impression. Professor Louis Nirenberg, who also knew Van from student days, was not put off by his French habits and manners: "He was just an extremely private person; *extremely* quiet. He never spoke up in class, and he didn't come to our parties, but I thought of him as a nice guy." Like Keller, Nirenberg noticed that "Van always had a very pretty girl in tow ... not always the same pretty girl either. He was an extremely attractive man. I could see what women saw in him."

Nirenberg was also favorably impressed by Van's Ph.D. thesis, and twenty-five years after its completion said he still found it to be "a nice piece of work."

Around the mathematics department, van Heijenoort was as close-mouthed about his political past as he was about his personal life. At first, he gave no hint of the character of his former life to explain the hiatus in his mathematical career, but inevitably, someone read or heard about his connection with Trotsky and rumors began to make the rounds. It is not at all surprising that this should have happened at NYU, for Comrade Van had recently played a part, however ineffective it may have been in his own view, in Trotskyist politics in New York City. Although his articles in *The Fourth International* had all been written under various pseudonyms, that sort of information has a way of leaking out. Even then, his colleagues almost never broached the subject of his close association with Trotsky. They did not ask why, what, where, and when about his past or try to discover where he stood at the moment. They didn't ask and he did not volunteer—with one exception.

In 1956, Van was a witness in the trial of Mark Zborowski, the KGB agent who the Trotskyites were convinced was in some way implicated in the mysterious death of Trotsky's son Liova. Van Heijenoort hoped that in the course of the proceedings precise facts about Liova's alleged murder would be revealed. Zborowski, in the United States since 1940, had continued his activities as a spy, and in the 1950s he was arrested, brought to trial in New York, and eventually convicted of perjury and sent to prison. Because van Heijenoort was sure that the trial would be reported in the press, he made an appointment to see Wallace Johns, the chairman of the mathematics department, at which time he quietly and carefully explained why Johns and others might find something about him in *The New York Times* the next day. At NYU, this was the first time Van ever said anything about his past without having been asked. Professor Johns was appropriately stunned.

But aside from the facts that Van felt obliged to reveal to his chairman, why was he so intensely close-mouthed? As van Heijenoort would have said, there were reasons on the political plane and reasons on the personal plane. In the 1940s, in the United States, both during the war and after, anyone having Trotskyite leanings was an outcast—"a political leper," in Van's words. The USSR had been the heroic wartime ally of the United States, and Stalin was pictured as the indomitable, stalwart leader. Although the Cold War would begin almost immediately following the end of the hot one, Trotsky and his followers did not by any means achieve a clean bill of health as Stalin lost his glow. During those years there was general embarrassment if not outright hostility towards people in radical left-wing politics and, except in very closed circles, the assumption (frequently correct) was that the people involved would prefer not to reveal their colors. To a certain extent this was as true of van Heijenoort as of anyone in his situation. But equally if not more important was van Heijenoort's extreme personal reserve, which absolutely no one who knew him ever failed to mention. The "very private person" aura he exuded inhibited everyone, and only the very young or the very brave broached the barrier. Some, like Sidney Morgenbesser, his colleague at Columbia University, said: "How could you invade that privacy? He was like a great wounded bird." Others, like Sidney Hook, saw him as cold, remote, and detached. But in whatever manner they read his external face, only rarely did anyone attempt to go behind the mask.

One brave and bright young man, Stefan Bauer-Mengelberg, did do so, very gingerly and tentatively at the outset. Bauer-Mengelberg, a philosophy undergraduate with a minor in mathematics, took a course in the foundations of mathematics given by Jean van Heijenoort in the spring of 1949. "It was to be my introduction to the whole question of the philosophy of mathematics, and there is no question that it stimulated my interest in logic," he said later. For at least a year before he enrolled in that class, Stefan had been aware of van Heijenoort, the strikingly handsome math instructor, who spoke with a heavy French accent and was a favorite among female students. Tongue in cheek, Mengelberg noted, "It was one of the principal recruiting routes in the math department because in some cases the attraction was strong enough to inspire young women to drop their old major subject and take up mathematics." The silent Frenchman's amazing popularity with his female students soon became part of the folklore of the math department, but Bauer-Mengelberg went one step beyond what was common knowledge to learn something else.

He had had occasion to look into Trotsky's biography of Stalin, published posthumously, where he had noticed an acknowledgement to a M. van *Heijinoort* for his help in preparing the book for publication. Bauer-Mengelberg, who fled from Germany in 1939 when he was twelve, had more than a casual interest in politics. Intensely curious about the mysterious instructor, he simply went up to him one day as he was standing in Washington Square, and with no preliminaries, as he later related,

> I asked him whether a member of his family spelled his name with 'j i n.' He asked me why I wanted to know. I explained about the acknowledgement in the book and then Van said, "Oh, it is me."
>
> I said, "How come the initial 'M'?" And he said, "Oh, that is for *Monsieur*." "And the 'j i n'?" I asked.
>
> He said, with a smile, "Oh, you know, American publishers—a misspelling."
>
> He was friendly, but I don't think we discussed the matter further. In those days, he was such a private person that if you met him in the corridors and said hello, he would blush before he said hello back.

How much of van Heijenoort's reclusive if not secretive behavior at NYU was deliberately designed to keep a low profile and conceal his political involvement and how much of it was due to his characteristic reserve and timidity is obviously an unanswerable question, but it is safe to say that these two elements, augmented by a healthy dose of general anxiety and uncertainty about his ability and status as a mathematician, as well as his insecure status as a resident alien in the United States, rendered him silent much of the time and kept his colleagues at bay. If they knew anything about his political life, they rarely brought it up and certainly never confronted him with it, and thus they were unaware that he had any internal conflict on that subject.

So the mathematics department of New York University was van Heijenoort's safe-house. He took refuge in his research, and with his well-honed talent for concentration he was successful in convincing others, and even himself part of the time, that his major passions were geometry and convex surfaces and his major battles involved the proof of mathematical theorems. In truth, that was only one, the newest and clearest, compartment in his life. The rest was a muddle—of ideology and emotion, of attachments and estrangements, of old loyalties and betrayed expectations—which would not fit into compartments and took van Heijenoort a good part of the rest of his life to sort out, if indeed he ever did.

The reality of van Heijenoort's profound experience of alienation in New York and his estrangement from the leadership of the Fourth International drove him to a reevaluation of all that he had held holy. Rereading Marx and Engels and reexamining the mathematics Engels invoked to validate the theory of dialectical materialism, he found the attempted justification weak, shallow, and already fifty years out of date in 1848, when the *Manifesto* was written. But finally, and most significantly, the culminating factor in his rejection of Marxism was the postwar revelation of what had occurred in the Soviet detention camps: the sheer, unbelievable numbers of people imprisoned and annihilated. Facing the horrifying facts, van Heijenoort came to the conclusion that it was not Stalin alone but the Marxist system as a whole that had created the potential for the tragedy, and that had Trotsky been in power, he might have committed similar atrocities. Nor could he avoid one further step in his reasoning: if Trotsky might have committed atrocities, what would the loyal Comrade Van, eager to do whatever his leader thought right and necessary, have done? The answer terrified him.

After a long and tortured process during which he turned himself inside out and outside in, weighing and measuring carefully and rationally, as was his wont, all the relevant parts and pieces of the ideas he had once held to be true, van Heijenoort wrote a long article for the March 1948 issue of *Partisan Review* ("the best literary magazine in America," according to *The New York Times Book Review*).

To mark the hundredth anniversary of the publication of the *Communist Manifesto, Partisan Review* published "A Century's Balance Sheet" by Jean van Heijenoort written under his Jean Vannier pseudonym. Sandwiched between "The World is a Wedding," a short story by Delmore Schwartz, and "Atlas on Grass," a poem by Vernon Watkins, was Comrade Van's public repudiation of Marxism as a viable political doctrine. The article took a scientist's analytical stance and examined the *Manifesto*'s fundamental hypothesis, "that the proletariat can and must take the fate of society into its hands, snatch it from the catastrophe into which the bourgeoisie is leading it, and guide it toward communism... To Marxist commentators this premise seems as solidly based, or nearly so as the laws of motion governing the heavenly bodies."

Van Heijenoort, now outside the Marxist circle, insisted upon more rigorous scientific proof:

> Without pausing to examine just what the modern physicist considers as truth in this domain, it is enough to observe that Neptune still follows the orbit marked out for it by Leverrier, whereas, unhappily, the course of the proletariat has been increasingly erratic. So much so that it is impossible today to shrug off the necessity of a systematic scrutiny of Marx's fundamental hypothesis.

And, after just such a "systematic scrutiny" and the totting up of plusses and minuses, "A Century's Balance Sheet" concluded almost as it began:

> Nothing is gained by saying: let us hold fast to the fundamental hypothesis as long as nothing better is proposed. To receive answers to questions, one must first raise them. It is only by openly facing the difficulties of the fundamental hypothesis that a step to their solution can be taken. It would be as idle to shut one's eyes on them as to turn one's back on politics. In either case, the problem is still there. Only by a rational and methodical

> scrutiny of the lessons of the past and of present possibilities will we be enabled to work effectively toward preparing for a future. Whoever is content at this late date to go on repeating the basic hypothesis without advancing some new and decisive argument in its favor scarcely merits a hearing. Proposed solutions may well be widely divergent; only mutual criticism will make possible an intelligent choice. But in such an endeavor nothing can be held sacred—everything is called into question. Only after having been put through such a crucible could socialism conceivably re-emerge as a viable doctrine and plan of action.

To regular readers of *Partisan Review,* "Jean Vannier" 's article was hardly a bombshell. The crucible used to heat up Marxism was already well used and the description of the tests was a common theme in the pages of *PR*. But for the few who knew that Jean Vannier was Jean van Heijenoort, Trotsky's former secretary, the article *did* create something of a sensation. These were the readers Van worried about, former comrades or fellow-travelers who, he was certain, would label him a turncoat. From the direct experience of having been among the true believers, he knew the contempt that would be heaped on him, not only for what he said, but for where he chose to say it.

In *The New York Intellectuals,* a history of the decline of the anti-Stalinist radical left from the 1930s to the 1980s, published in 1987, Alan Wald, a professor of English literature and American culture, chronicles a long list of defectors from the Marxist faith. Among them is Jean van Heijenoort whom he describes as having made a "sudden and almost shameless volte-face." Van, who in later years habitually spent hours correcting errata in other people's books, would have had a fit of anger had he been alive to read this glib account by a man who knew almost nothing about him. van Heijenoort's "turnabout" was certainly not sudden, and "shameless" could not have been farther from the truth. The epigram for *A Century's Balance Sheet* was Hermann Sudermann's "Schuldig mussen wir werden, wenn wir wachsen wollen" (Guilt is inevitable if we wish to grow). Guilt and shame were Van's bosom companions; he embraced them to the point of speechlessness for having persisted in error for as long as he did. At the same time, he could not help but be distressed at giving up the ideals to which he had devoted his life; and even though he was finally sure of his decision, he was not impervious to the reactions of his former political comrades. He feared they would misunderstand; and forty years later, writing of Van's "sudden and shameless volte-face," Wald proves his fears were justified.

Van Heijenoort would have been equally disturbed by Wald's further assertion that he "chose as his last political act to publish his repudiation of Marxism in the *Partisan Review* under the pseudonym Jean Vannier. He then vanished from the political arena to begin a new life as professor of philosophy at Brandeis University." At times van Heijenoort may have wished he could have vanished, but he did not leave New York or NYU for another seventeen years, and while it is true that he left the political arena, in another sense he remained "married" to Trotsky until his own tragic death.

Only Bunny knew the torment her husband had gone through while writing his article for *Partisan Review* and making his conclusions public. She was aware of it more by the nature of his silence and withdrawal than by what he said, for he said almost nothing. She later recalled:

> He had some discussions with people still involved in the movement, but I think he mostly arrived at things by himself. I had the feeling he was trying to clear his mind and body, and he was searching his soul. I know it took a lot of courage for him to repudiate his old ideals, to admit that he had made a mistake, because when he did something—anything, he was *sure* it was the right thing to do. It must have been devastating to admit he was wrong about something so important to him. His reaction, at home, was mostly silence. He kept to himself, and he worked.

Although Wald suggests that it was somehow unseemly for van Heijenoort to publish his repudiation in *Partisan Review*, there was every reason for Van to do so. The editors, represented by William Phillips, had invited Van to write the article and for good reason. As noted above, his views were in tune, more or less, with those of the editorial board, many of whom were former Trotskyites or at least sympathizers. At its inception, in 1934, the magazine was sympathetic to Marxism but not formally allied with any political party on the left. For a brief period, *PR* supported the Soviet Union under Stalin's leadership. Then, as a consequence of the Moscow Trials, the purges, and the accusations against Trotsky, the magazine turned sharply away from Soviet Communism and, leaning towards the Trotskyite camp, described itself as being "revolutionary in tendency" but also "unequivocally independent." In their new statement of policy the editors wrote: "We think the cause of revolutionary literature is best served by a policy of no commitments to any political party." Nevertheless, many of the staff members had been active on the American Committee for the Defense of Leon Trotsky.

Soon after the Dewey Commission hearings in Mexico, Dwight Macdonald, one of the editors, had written to Trotsky inviting him to contribute articles to "an independent Marxist journal," and with that letter, written in July 1937, a lengthy correspondence had begun. The content of the letters ranged from what Trotsky might write, to whether he would or should associate himself with *Partisan Review,* to what the nature of such a journal ought or ought not to be. One of the results of this interchange had been *Partisan Review's* publication of the *Manifesto: Towards a Free Revolutionary Art,* the article that had caused so much tension between André Breton and Leon Trotsky during the surrealist's 1938 visit to Mexico. So Comrade Van, who had been in charge of correspondence, knew all about *Partisan Review* and knew who Dwight Macdonald was well before he arrived in New York in 1939. But even if the connection had not been established then, within the next few years it would have inevitably occurred because of van Heijenoort's friendship with Nicolas Calas, a regular contributor to *PR,* and more importantly, with André Breton, who, with his wife Jacqueline, had taken refuge in New York during World War II.

* * *

The Bretons' presence in New York from 1941 until 1946 meant a great deal to Van, who hungered after French companionship and news from France. His return to the university was still several years away, and, although he was still deeply involved in Trotskyist activities at that time, he was floundering and trying to get his personal and political bearings.

As in Mexico, Van went to greet the Bretons upon their arrival and concerned himself with their early needs as best he could. But since Van's own adjustment to New York was far from complete, their roles were often reversed. When it came to cultural life, Breton was the leader, the influential friend who introduced Van to interesting people and took him to interesting places. New York may not have been exotic and surreal *à la* Mexico, but it did have its attractions, and Breton, along with Calas and Abel, helped Van to discover them.

Already famous as the driving force of the surrealist movement, André Breton was the darling of the avant—and not so avant—garde. He was taken up by the *Voice of America* radio program, where Nicolas Calas was already employed, to read French texts on the broadcasts beamed to Occupied France, and he and Calas started the little magazine *VVV* (*Vie, Vie, Vie*). Breton was also a regular and important contributor to *Partisan Review* and through him van Heijenoort met and widened his circle of friends and acquaintances in literature and the fine arts. As a star in ascendance, Breton's glittering presence was much sought after and, happily, some of his luster rubbed off onto his younger comrade.

In art, music, and literature, just as in the domain of science, one of the well-known beneficial side effects of World War II was the infusion of new ideas and perspectives provided by the influx of the talented Europeans who came to the United States as refugees. From Einstein to Lévi-Strauss to Billy Wilder, the exiles enhanced and changed the cultural, intellectual, artistic, and scientific climate. Although Van was almost unremittingly unhappy in those years, he later acknowledged, grudgingly, that it was an exciting time. "Odd, troubled, strange," he would say. The war was going on in Europe, his country was occupied, he was poor, his life in New York was harsh, and his "job" brought him little satisfaction. But in the midst of what seemed to him the worst of times, he was enjoying the company of famous surrealists and abstract expressionists, well-known writers and critics, and the coterie that followed in their wake.

On one occasion, Breton took Van and Bunny to a party given by Peggy Guggenheim and Max Ernst that Van, in his later account, acknowledged was a stellar evening:

> There were easily seventy or eighty guests at their huge, swanky apartment. I didn't know exactly who these people were or where they were coming from. Of course, there were also many people I knew: Jacqueline (Breton's wife), Nicolas Calas, Lionel Abel, Arshile Gorky, the painter, and Yves Tanguy, who I liked very much.

Part of the excitement was a showing of the documentary, *Ten Days That Shook The World,* based upon the John Reed book, and narrated by Max Eastman. As soon as the film began, the sound machine broke down, whereupon into the silence stepped Eastman himself, who was conveniently present. "He improvised the text following the movie as it went along," van Heijenoort remembered, "and it was better than what was in the movie; he told more and the way he told it was quite charming."

The *haut monde* that invited the Bretons to their parties in the city also invited them to their weekend retreats in the country. Although, in the 1940s, East Hampton and its neighboring communities on Long Island were not the "in" second-home meccas they would become in the decades to follow, a few wealthy families had already established a beachhead on the rural eastern end of the Island early in the century. The invitations offered an escape for Breton, who for all that he was lionized and flattered, was basically unhappy in New York City for some of the same reasons as van Heijenoort. "A city for giants" he called it, and he was only an average-sized man. He found the buildings too tall, too grim; the canyons between, too narrow. He missed the architecture and the ambiance of Paris. When the war ended, he returned to France as soon as he could—one of the first refugees to do so. But until then, the outer edge of Long Island was his refuge from Manhattan and when the opportunity presented itself, he would go, bringing Van and as many other Frenchmen as he could along with him.

To be sure, Breton did not always have an invitation to a house party in one of the Hamptons or Sag Harbor. Then, the French expatriates and their spouses, including Van and Bunny, would get together and rent a house for a week or two and go out on their own. Van felt happy and overcome with nostalgia. "When I first went out there," he said later, "there was nothing but potato farms. It reminded me of the countryside around Creil; it was the closest thing I could find to being in France." And it was the truest kind of nostalgia, for it invoked a peace he had seldom known as a child.

In the city, the closest one could come to France was the quasi-Parisian atmosphere of Greenwich Village. Van Heijenoort was content to live there with Bunny and Jeannot after their return from New England in 1944, and he continued to live in the Village until he left New York in 1965. In and around Washington Square, surrealists, socialists, and Trotskyites of varying stripes and hues mingled and formed a set. And after 1945, when Van entered NYU, even as others perceived him as the "oddball loner," at least in some degree, he felt a sense of belonging in the academic community, too.

While Van worked at mathematics as a graduate student and a teaching assistant, Loretta took classes at the Art Students' League and had jobs doing theatrical makeup. Jeannot was a junior high school student, riding a bike, playing tennis, and merging into the American melting pot. Except on Sundays, everything revolved around Van's work, whether it was mathematics or politics. Often he was out, but if he was at home, after lunch or dinner, he would disappear into his study. As Jeannot later remembered and Loretta concurred, there was no chit-chat or small talk at home, but rather an unbroken silence which frustrated them both.

Sundays, however, were less serious, "the best day of the week," according to Jeannot:

> We would get up late and read *The New York Times* that I had fetched. Bunny would do French cooking and pastry, and my father also made some of his *"specialités"*: omelette flambé, crêpes, and mussels. Sometimes people would come for dinner, or I would go with my father and Bunny to other people's homes.

Sunday had always been Jean van Heijenoort's "best day" as well. As a child it had been the day he saw his mother, and as a boarding student at Clermont and the Lycée Saint-Louis, it had been the day of freedom. The old feeling of release stayed with him, and the image Jeannot would retain from the special days, in contrast to the general picture, was of a more open and relaxed father who even joked and teased.

In those years, Van and Bunny also had some social life centered around the unofficial "Eighth Street Club" or "Artists Club," a floating group of artists and intellectuals who got together in a rented loft in an old brick building on Eighth Street for informal lectures, discussions, and parties. Naturally, the focus of most of the painters' talk was aesthetics; but on many occasions writers associated in one way or another with *Partisan Review*, including the art critics Clement Greenberg and Harold Rosenberg, would drift in. Thus, literary and political matters also worked their way onto the agenda. Typically, as an evening at the club wore on, the atmosphere would become less and less formal, and the party would begin. Later, people drifted off to neighborhood cafés and bars, usually the White Horse or the Cedar Tavern.

Among the Eighth Street crowd of painters, critics, and literati, van Heijenoort was unique. There, in contrast to the halls of NYU, everyone knew he had been Trotsky's right-hand man and bodyguard, but they were very surprised to learn he was also a mathematician. It would be hard to say which of his attributes gained him more respect in that group. Being the bodyguard of a "wanted" man requires a kind of physical bravery that few people have put to the test, but being a mathematician requires a kind of mental agility that most people—intellectuals included—are sure is beyond them. In addition, it soon became evident that the quiet Frenchman with the Dutch name also knew something about art, science, history, literature, and language.

In due course, van Heijenoort accepted an invitation to give one of the weekly talks at "the club." He chose mathematics and its relation to abstract art as his topic and discussed different types of geometry: how they might be exemplified in painting or used to interpret paintings. His listeners included some of the most active Eighth Street artists: Elaine and Willem de Kooning, Franz Kline, Jackson Pollock and his wife, Lee Krasner, all of whom would become famous as leaders of the abstract expressionist "school." Asked many years later whether he thought his talk had any effect upon them, Van Heijenoort demurred. Instead he recalled a polemic delivered by Heinrich Blücher, Hannah Arendt's husband, entitled "Kitsch in Art," and Willem de Kooning's whispered question which he insisted was serious: "Tell me, Van, who is that guy Kitsch that he keeps talking about? I've never seen any of his paintings."

Van Heijenoort was stimulated and entertained by the cultural and social atmosphere of the Artists Club. Even though he was deeply involved with his mathematical work and constantly under pressure to solve new problems, he allowed himself to spend at least one evening a week with the group. He was particularly attracted to the painters and approved of their artistic experiments. But at the same time, he expressed scorn for the social value of their work. One day a colleague at NYU, knowing of his participation in the Artists Club parties and discussions, expressed a wish that there could be a similar kind of group at NYU. "It must be wonderful," she said. Stefan Bauer-Mengelberg, who was present at the conversation remembered Van's response: "Wonderful? Well, they are plain stupid. They think that painting is metaphysics; and that by painting they will save the world. Can you imagine? Painting, in New York?"

* * *

By 1948, Van had given up trying to save the world. Although the process of weaning himself from Trotsky and Marxism had hurt, he knew he had done the right thing. At the same time he feared that in abandoning his long and passionate commitment to his former ideas and ideals he had lost something he might never recover:

> I felt like I had been pushed over, and when I got up, I was off balance and could not regain my equilibrium. I could never have my two feet firmly and securely on the ground again.

And as late as 1985, van Heijenoort described a recurring dream that he had had for as long as he could remember:

> I am riding a horse down the main street of my town in France. It is war time. All of a sudden the horse collapses; his legs just give out, but instead of falling in the usual way, all four of his legs are splayed out to the side. I sit there on the horse and look at those legs and wonder how I am ever going to get up?

Chapter Twelve

# The House That Jean Built

New York: 1945–1956

*Father and son in New York, just before Jeannot's first return to France. 1946.*

"Oh, I am again on zee deportation list."

JEAN VAN HEIJENOORT *to Stefan Bauer-Mengelberg*

WHEN WORLD WAR II ended in the summer of 1945 and it was again possible for civilians to travel from the United States to Europe, there was nothing the expatriate Jean van Heijenoort yearned to do more than jump on the first available ship and return to France. It had been nine years since he had seen his mother, his family, his friends, and the places he thought of as home. With envy he had watched André Breton happily pack up and leave the "City of Giants" for the City of Light. But this was not to be one of Van's "I didn't think twice" decisions. He was thirty-three, not twenty, he was trying to build a new life, and he did not want to close off his options. Because of his status as an alien, if he left the United States, he had no assurance that he would be allowed to return. The immigration authorities had made that clear, and before long it also became clear that even if he chose *not* to leave, he was a likely candidate for deportation. Weighing the pros and cons, van Heijenoort decided to stay in New York—to wait and see what would happen. It was not a short wait and a great deal happened.

In light of the rigorous scrutiny undergone by later generations of immigrants, it may seem surprising that van Heijenoort encountered no difficulty in 1939 when he entered the United States as Loretta Guyer's husband. Possessed of an entry visa and all other necessary documents, he assumed (correctly, as it turned out) that the officials of the Immigration and Naturalization Service knew nothing about his political history and his connection with Leon Trotsky. As ever, he had answered the questions he was asked in his typical minimalist fashion. For example, when asked his occupation, he had answered, "translator," omitting the name of his employer as well as the nature of work. In 1940, he filed his intention to become a United States citizen and he also registered for the draft with the Selective Service Agency which classified him as 3A (married).

For the next few years, Van still had no indication that his past had come to any government official's attention or that he stood out in any way. Using pseudonyms and moving frequently, he took many precautions to remain unnoticed. Between 1939 and 1944 he relocated no less than ten times, in New York City, Baltimore, Philadelphia, upstate New York, the Boston area, and even Tucson, Arizona. But by 1945 he had become aware that things had changed and, as he put it later, "the FBI, the INS, the Army, the Navy—all of them—knew everything about me."

In fact, they did not know *"everything,"* but the government had indeed "developed," (in the FBI's terminology) a rough picture of van Heijenoort's past, and they had done so earlier than he realized. In August 1942, in the midst of the war, John Edgar Hoover, then director of the FBI, signed the following directive to the special agent in charge of the New York office "RE: Señor M. Loris":

> I am enclosing herewith photostatic copies of a submission sheet received from the National Censorship [Bureau] concerning a communication dated April 15, 1942, written by an individual in Mexico to the above captioned subject [M. Loris] at 116 University Place, New York, New York. In view of the address of the subject [Loris (van Heijenoort) used Socialist Workers party headquarters as a mailing address] and the contents of the letter directed to him, it appears that the subject is a Trotskyite Communist.
>
> You are requested to conduct an appropriate examination to develop the background and activities of the subject with reference to his possible detention.

The New York office did not leap to the task; instead, the case was deferred until staff was available. Sixteen months later, Hoover again wrote to the New York office "RE: Marc Loris." Noting that a review of the file "fails to reflect that your office has conducted any investigation concerning this subject's activities ... ," he gave orders:

> You are instructed to assign this matter immediately and to submit an investigative report in the immediate future containing information regarding the subject's background and the extent of his present activities, in so far as they affect his attempt to carry on relations for the Fourth International in this country and with other individuals outside of this country. This matter should be considered of importance and preference should be given to this investigation.

Chided by Washington, the New York office applied itself. Combining reports of their technical and physical surveillance of "the subject," the Socialist Workers party headquarters, and copies of further correspondence between M. Loris and various members of the Fourth International in Mexico, France, England, Italy, and South America, the beginnings of a dossier were assembled. Rather quickly, important pieces of the puzzle interlocked: a sketch emerged, and then a New York agent enlarged the picture with the discovery that "Marc Loris is believed to be identical with one Jean van Heijenoort. ... It is thought possible that his employer may have been LEON TROTSKY."

Dramatic news. Now the Loris/van Heijenoort reports carried the heading, "Key Figure." According to the Washington Bureau, "It would appear that Loris is unquestionably the leading functionary of the Fourth International movement in the country and perhaps in the Western Hemisphere ... It is believed that his activities warrant a comprehensive and vigorous investigation."

In concert with the Immigration and Naturalization Service, between 1942 and 1981 the Federal Bureau of Investigation was to compile a 556 page investigative report on Jean Louis Maxime van Heijenoort/John Louis van Heijenoort/Marc Loris/Daniel Logan, [et cetera]. The document, which remained classified until 1988, contains information from many sources, including that provided by the subject himself when he entered the United States, when he applied for citizenship, and when he registered for the draft. There are also transcripts and summaries of his many interviews with the INS and the FBI, and reports of the agents who followed him for several years and gathered information from his employers, his colleagues, his landlords; and there are copies of personal as well as political letters intercepted and submitted by the Censorship Bureau.

Ironically, much of the material was gathered after van Heijenoort had ceased his activity with the Fourth International and his loose connection with the Socialist Workers party. While the FBI was busy catching up on the old Comrade Van and considering the possibility of deportation proceedings, the new Van had become a graduate student at NYU, frantically catching up on the literature of mathematics and philosophy he had missed during his years of total absorption in politics.

But eventually the US authorities were *au courant*. Certainly by 1948, after Jean "Vannier" published his renunciation of Marxism in *Partisan Review*, it was clear to even the most skeptical that the "key figure" was no longer a security risk and the designation itself inapt. The INS agent who had interviewed van Heijenoort at length and would continue to do so for the next few years appears to have succumbed to his charm, describing him as "a man of brilliant intellectual accomplishment ... frank, engaging, and sincere ... who is aware that his admissions leave him vulnerable to deportation proceedings." The New York FBI agent, on the other hand, was less beguiled. To J. Edgar Hoover, he wrote:

> It is true that he [van Heijenoort] has been to some extent cooperative ... as would be expected of one in quest of American citizenship. However ... he describes his activities in the United States as mere "sympathy" with the American Trotskyists, whereas in fact there is strong reason to believe that he was, during the war years, a guiding genius of the world-wide Trotskyist movement, or at least a coordinator thereof.

In spite of a tacit, if grudging acknowledgment that van Heijenoort was not conspiring against the United States, his request to become a citizen was not to be granted for another ten years, and for much of that time the threat of deportation remained a specter, waxing and waning with the political mood of the country. In the early 1950s, during the heyday of Senator Joseph McCarthy, the notorious witch-hunter of leftists and even their casual friends, socialists and liberals of paler shades of pink than van Heijenoort were at risk. In those days, Stefan Bauer-Mengelberg, van Heijenoort's student, who by then was also teaching mathematics courses, would find Van sitting on a bench in Washington Square, apparently deep in thought.

"I would pass him on the way to my evening class," Stefan later recalled.

"How are you?" I would ask, and he'd answer, "Ohhh, I am again on zee deportation list. The immigration man comes to see me all the time."

Alternately brandishing the stick of deportation and dangling the carrot of citizenship, the FBI sought to enlist van Heijenoort's cooperation in their search for evidence that the Socialist Workers party and the Fourth International advocated the violent overthrow of the United States government. They also hoped that he would name individuals involved in those organizations. In response to requests from both the FBI and the INS van Heijenoort agreed to tell what he knew, with the qualification, duly noted in the FBI report, "that although he had abandoned Marxism he could not turn on old friends towards whom he felt a certain personal loyalty. Moreover he said such information as he could give is now old and largely unimportant and the passage of time has faded his memory."

Van Heijenoort later said, and it appears to be true, that he never told the FBI of anything or anyone they didn't already know, except in the cases of Mark Zborowski and Jack and Robert Soble who were accused of espionage in 1956. In these instances, he was more than eager to provide information and even volunteered to testify in the Zborowski trial because he had an old score to settle. He was convinced that Zborowski had played a leading role in the theft of Trotsky's papers in Paris and, worse yet, had been involved in Liova Sedov's mysterious death. In Van's view, Zborowski and the Soble brothers were Stalinist agents who had masqueraded as supporters of Trotsky and in various ways had subverted the cause to which he had once devoted himself. The ex-Comrade Van's politics may have changed, but his feelings about such people never did.

Although in the end, the offer to testify was not taken up, the FBI did accept Van's proposal to review the Trotsky Archive at Harvard University and to interpret and translate the relevant papers for two agents who accompanied him to Cambridge for this purpose. Professor van Heijenoort's expenses were paid by the Bureau. This spirit of cooperation seems to have been one of the factors that finally opened the door to citizenship, although van Heijenoort himself played down this explanation, saying when asked that he didn't know what had done the trick unless it was simply the passage of time and the merits of his case. This response is one of the few occasions where Van appears to have been disingenuous. Perhaps he was embarrassed about his "cooperation" with the FBI at a time when some of his colleagues would have considered any such alliance a bargain with the devil. But van Heijenoort also neglected to mention letters written on his behalf by various influential friends of friends, including John David Lodge, then a Massachusetts congressman.

Obviously, during the long years of uncertainty about his status, it was unsettling for van Heijenoort not to know whether he would be permitted to stay in the United States. Although in retrospect he claimed he would not have been upset if he had been deported to France, he was not naive about his prospects there at that time. Without knowing exactly what the situation was in the universities in France, he suspected that in the aftermath of the war, money for higher education would be limited and the educational system as a whole, disorganized. Even if the universities did get their programs underway, there would not be the kind of financial support that he was assured of as a teaching assistant at NYU.

There was also the matter of military service in France. Van Heijenoort had never done his obligatory tour of duty, and even though he had been absent since 1937, the military had never lost track of him. Hélène had sent word that *les gendarmes* had come to the house in Creil asking where Jean van Heijenoort was. Van feared that if he were to return to France he would be obliged to spend two years in the army, a prospect that chilled his every thought of repatriation.

Nevertheless, all the sound, practical reasons did not quell Van's desire to see his people and his country, and between

1945 and 1958, the question of whether to risk a return nagged at him. He talked about going, vacillated, claimed he did not really care if he was deported because he never had decided he really wanted to stay, but in the end, he did not set foot outside the United States until he had become a citizen and was certain that the option to return was his. Nor did he set foot inside France before his old comrade in Paris, Gérard Rosenthal, a lawyer, had settled the problem of his military service.

In the interim, Van sent his wife and child as his surrogates: Bunny, to tell all that had transpired in the decade he had been gone, or what she knew of it, for there was much that Jean had not confided to her; and Jeannot, to have a reunion with Gaby, his mother, and Grandmother Hélène.

They went in the summer of 1946. For Bunny, a most willing representative, the journey was an adventure and a challenge. She was going to Europe for the first time, to meet people she did not know, including her mother-in-law and her husband's ex-wife, who were very close. They did not speak English, and Bunny spoke very little French. In this delicate situation, loaded with potential for unpleasantness if not disaster, Bunny used all the considerable charm and finesse at her command to thread her way through the intricacies.

Their reception was warm, but for Jeannot, now eleven years old, the trip and visit were a mixed blessing–a reminder of his double life and the multiple ties and loyalties he was obliged to keep track of and deal with. After five years in the States, he had forgotten French and scarcely remembered his "real" mother and grandmother. He thought of himself as American, and confusing though it was, he quite naturally regarded Bunny as his mother because she had acted in that capacity since he was six years old and had provided the most stable atmosphere he had known in his peripatetic life.

The "Americans" spent the whole summer in France, mostly in Hélène's house in Nogent-sur-Oise, on the edge of Creil, where Bunny became familiar with the territory of her husband's youth. Jean had told her all about his brilliant mother, and Bunny, the savvy New Yorker, agreed: her mother-in-law was

*Bunny and Jeannot, ready to embark. 1946.*

"sharp as a tack." In spite of the language barrier, Hélène's warmth, intelligence, and tact came across. And then, there was her magnificent cooking.

"I loved her from the moment I met her," Bunny later said. "Right from the start, I called her *Maman*, and she called me Bunny. We were very close ... like family."

While Bunny told about Van's life in Mexico and New York, Hélène recounted the older stories of her son's boyhood, the bitter tragedy of his father's death, and the tales of hardship from one war to the next when the hated *Boche* had invaded the town. Once more they had become a family of women without men, for Georges Doré, Hélène's second husband, had died in 1940, leaving Hélène, her daughter Paulette, her mother Adèle, and the two sisters and nieces to fend for themselves. They had survived the war, but in 1946 the suffering they had endured was still evident.

Long thwarted by her husband's silences, Bunny was an eager and appreciative listener. Encouraged by her daughter-in-law, Hélène went into loving detail about her son's childhood tantrums, illnesses, and miseries. Naturally, she also celebrated his brilliance, achievements, and honors as a student. It became clear to Bunny that Jean had been Hélène's pride, joy, and consolation, the one member of the family who had been given the opportunity to fulfill his potential. Finally, Hélène told of her chagrin when Jean had gone off to join Trotsky, giving up everything he had worked so long and hard to achieve. She had lived in fear that one day a customer in their Café-Restaurant "A La Maison Doré" in Saint-Quentin might see his picture or his name in the paper, next to that of Leon Trotsky.

Although these memories and worries were still vivid, they were stories that had already passed into legend in the family. But there were more recent tragedies for Jeannot and Bunny to hear, stories of wartime hardship and death. Saddest of all was the misfortune that had befallen Gabrielle Brausch after she sent Jeannot off to safety in America in 1941. She had returned to Marseilles to continue her political work, along with her companion, Maurice Segal, but within a few months she was arrested and put into prison, first in Marseilles and later in Rennes.

The inadvertent agent of Gaby's imprisonment was a Czech who had flown from England to France and parachuted into territory near Marseilles, in order to bring information in aid of the Resistance movement. The operation was a complete failure: the man was captured, and among his effects was a list of names and addresses, presumably of people he was supposed to contact. One of the names on the list was Gabrielle Brausch. Although she insisted that she did not know who the man was or how he had obtained her name, she was charged and convicted as a co-conspirator. Van Heijenoort, in recounting the story later, never failed to wince at the gross carelessness: "If he couldn't remember, at least he could have used pseudonyms." Fortunately, the parachutist was caught by the Vichy French who were then the nominal authority in Marseilles. As a result, Gaby was incarcerated in French prisons rather than a German concentration camp, which would have been her fate had the same event occurred in occupied France. In jail, Gaby became physically sick and emotionally disheartened, but, all things considered, she may have been safer in prison than out.

Maurice, her companion, had a worse fate. As a member of the Resistance, he was captured by the French and imprisoned at Le Puy-en-Velay. He and other prisoners escaped in 1943

*Jeannot and his grandmother, Maman Hélène in Nogent. 1946.*

and took refuge among the *Maquis* in the region of the Haute-Loire. There, according to the historian Pierre Broué, an expert on the Trotskyist movement, he was almost certainly murdered by a Stalinist, one of his own comrades in the Resistance, who was under orders to wipe out the dissident Trotskyites whenever and wherever the opportunity presented itself. After her release from prison, Gaby went to Le Puy to investigate Segal's disappearance, and later tried in vain to find witnesses who would testify to what had happened. "She was the only one who had the courage to do that," Broué said later.

With the liberation of France in 1945, Gaby returned to Paris depressed and in poor health. She found work, but distanced herself from politics and her old comrades. She did, however, remain close to her ex-husband's mother, the wonderful *Maman* Hélène. When Bunny arrived with Jeannot in 1946, she, of course, met Gaby who was also part of the family and a regular visitor in Creil. Surprisingly, there was little jealousy between them although later there would be some conflict over Jeannot. Instead, bound by love and concern for "their boy," Gaby and Bunny, the two "mothers," became good friends and spent a good deal of time together.

As a result they sometimes found themselves in odd social situations. Bunny later remembered a day when she, *Maman*, Gaby, and Jeannot, went out walking and chanced to meet *Maman's* distant cousin—a man she hadn't seen in years—who had not met either of the two wives. Hélène proceeded to make introductions saying, "This is Jean's wife, Gaby, and this is Jean's wife, Bunny, and this ... is their son, my grandchild, Jeannot." Gaby and Bunny looked at each other, turned to the cousin and said hello—politely as they could—and then walked away, giggling. After the cousin left, *Maman* began to scold her daughters-in-law for their bad manners. Protesting, they pointed out the absurdity of what she had just said. Hélène answered, "Well, what else could I have said?", and she, too, burst out laughing.

The least comfortable member of the party was its star attraction, young Jeannot, who, while being loved to death by everyone in sight—mothers, grandmothers, aunts, and cousins—felt totally out of place. Unsure of where and to whom he belonged, he had no regrets when the summer came to an end and it was time to go back to the familiar territory of Greenwich Village.

Jeannot lived in the United States for two more years, until he was thirteen. Then, in 1948, he returned, or to be more precise, was returned, to France. Had he been given a choice, he would not have elected to go. "I wanted to stay in the States," he said later. "I felt I had nothing to do with France any more; I had left the people there when I was six and I didn't speak a word of French. I thought of myself as an American." The choice, however, was not his and it was not Bunny's. In rare agreement, and in contrast to their political bickering, it was Gaby and Van who made the decision.

The most compelling argument was Gaby's: she wanted her son. She had come out of jail with a troubled heart, physically weakened as well as emotionally exhausted. Her companion Maurice was dead, she had formed no new personal attachment, and disillusion had caused her to abandon political activity. An emptiness had come into her life, and she needed her son to help fill the void. Van Heijenoort saw the justice in this and also had some reasons of his own. Jeannot was ready for high school, and it was very important to Van that his son have an excellent education. In his opinion, the best place for that was an academic *lycée* in Paris and not a public high school in New York, although he cannot have known much about New York high schools. In spite of his talk of the cruelty and absurdity of the discipline and regimentation in force during his own schooling, when it came to making a choice between intellectual rigor and emotional or physical comfort, he chose the former for his son—as he probably would have done for himself.

Van's personal motivations were less clear but also a factor. By 1948, his marriage to Bunny was strained to the breaking point; it was evident to both of them that a separation was imminent, and in fact they did separate temporarily, although not finally, at that time. Their marital problems did not affect Bunny's love for Jeannot—perhaps her feelings may even have been intensified. She wanted her boy to remain with her, particularly since he himself did not want to leave. She pleaded and pressed her case, arguing that in effect, he was more *her* son than anyone else's and would be happier with her, but she could not persuade Van to change his mind. In the days when he had been Comrade Van, Jeannot's welfare had had to take a back seat to his commitment to Trotsky. Now that those global considerations were behind him, van Heijenoort's major commitment was to mathematics and scholarship, but his emotional involvement with his son had not changed very much. To be sure, he felt a strong parental responsibility towards him, he had provided for his safety during the war; but if he had a deep attachment and wished for his son to be physically close to him, it was well hidden. It would seem that when he came to the conclusion that it was right for Jeannot to leave New York and to be in Paris with Gaby, seeing the boy go did not hurt him as much as it did Bunny or Jeannot.

So Jeannot crossed the Atlantic once more and went about the business of adapting to French ways and to the realities of postwar France. Besides the personal difficulties of adjustment, there were also the hard facts of limited money and scarcity of goods. Food was still being rationed: "We had to have coupons to buy bread," Jeannot later recalled. "Life was not easy for me and my mother."

Nevertheless Jeannot did readjust, and, in retrospect, he seemed to have few regrets. He learned to speak French again in nine months, and after passing the examinations with high marks, he was admitted to the Lycée Louis le Grand, the other great academic preparatory school. At Louis le Grand, he said later, "The work was hard, the hours were long, there was very little time for fooling around. Every day I would study until one or two in the morning—very different from an American high school." But, he was not an interne as his father had been at the Lycée Saint-Louis; he lived at home with his mother, and when school was out, *he* was out. For those hours at least, he was not a prisoner locked in his institution, but rather, one of the *externe* boys his father had envied two decades earlier.

The mature Jeannot made much of that distinction. In his view, it explained many of the differences between him and his father. "I was never sent to one of those terrible boarding schools with that artificial atmosphere where you get a good education but no preparation for life." Although from infancy Jeannot had been shuttled from place to place, and rarely lived with his father and mother, he always lived with family. In Paris, Creil, Mexico, Marseilles, Lisbon, New York, Massachusetts—he was "at home" with at least one nurturing woman. In contrast to his father, who unequivocally declared himself to have been "a disturbed child," Jeannot described himself as "a happy, easy-going kid" who did the best he could to adapt to any situation he was in.

There is one more obvious difference in the life history of father and son. However absent and detached Jean van Heijenoort *père* may have been, he was not dead, and he was someone of substance, in at least two meanings of the word. Whatever Jeannot, as a boy and later as a man, may have felt about his father personally, or about his father's politics, he ended by having profound respect for his father's unflagging commitment to use his time on earth doing what he thought were important things with his life, both for himself and others.

* * *

Van Heijenoort's constant inner turmoil during the decade between 1940 and 1950 was expressed by being unexpressed, or, one should say, unspoken, and this reticence had a disastrous effect upon his relations with Bunny. Without taking her into his confidence and without any discussion of his consuming worries even on a superficial level, he had expected her to understand his deep silences and to make allowances for his depressed state of mind. When he first came to New York as Trotsky's representative to the Fourth International, he had asked her not to do political work, saying it was potentially dangerous and that it might compromise his position as international secretary. It is not unlikely that Van also had in mind the conflict that had developed between Gaby and him during the factional fights within the movement in France, as well as the final, disastrous blow-up between Gaby and Natalia Trotsky in Mexico. In any case, Bunny, a more flexible comrade than Gabrielle Brausch, agreed to be inactive.

Although Trotsky was no longer alive and the country was not France or Mexico, the United States was equally fertile territory for disagreement, and hostilities between the Trotskyite factions flourished. Groups split, attitudes were condemned, new alliances and allegiances were formed. All of this Bunny knew from a distance, but she was never directly involved, as Gaby had been. Her husband wanted her to be the good wife, to go along with whatever position he took, and to take care of his son. She did all of those things gladly, but tension grew between them nonetheless, stimulated by the circumstances of his defection, and by his detachment from what his son would call "ordinary life." Where once politics and "the cause" had been the motivating force for all his thought and action, mathematics and philosophy and cerebral pursuits were now his central occupation, preoccupation, and *raison d'être*. The rest of life's business—according to the impression he gave his family, in attitude more than words—was too trivial, and he did not have time for banalities. His job now was to catch up and make up for thirteen lost years of mathematics and to solve a few unsolved problems.

* * *

As his colleagues at NYU and almost everyone else who knew him never failed to mention, Jean van Heijenoort was magnetically attractive to many women. When he was in the mood, he could be charming, attentive and even talkative, and the mood did come over him when he found someone who pleased him. As a teaching assistant and young faculty member at NYU, there was ample opportunity for him to meet young women, to invite them out for coffee and let matters proceed from there. More than once, before Van and Bunny finally divorced, he had become involved with one of his students and had separated from his wife. At those times, Bunny, too, made new romantic attachments; but if it looked as if she was more than casually interested in her new boyfriend and ready to begin a new relationship, Van would reappear. Always, at what seemed to be the final breaking point, he would say he wanted to come back to her. He would make promises, she would succumb to his charms, there would be a wonderful, brief reconciliation, and then the story would repeat.

The on again, off again torment went on for a few years until there was an "off" period during which Bunny met someone she thought she loved, and she accepted his proposal of marriage. Finally, she had decided, as she later recounted,

> ... that I was definitely going to divorce Jean and marry that man. When Jean found out about it he became insanely jealous; he followed me around; he hounded me and begged me to reconcile with him, but this time, I refused. My boyfriend and I were actually on our way to the airport when I said that I had to stop at my apartment and pick up something. My friend begged me not to do it. Then he said *he* would go up and get it for me, but I insisted that I had to do it. Jean was there—of course—and he went into hysterics when I told him I was leaving. He looked absolutely terrible, like he had lost twenty pounds. He had stopped eating. He said he couldn't live without me, that he would kill himself if I left, that he would do anything for me, that if I wanted a child, he would agree ... That had been one of our differences. I had never seen him act so crazy.

Bunny's description calls to mind the scenes from van Heijenoort's childhood when the six-year-old Jean staged tantrums for his mother as she prepared to return to her job at the Hôtel du Chemin de Fer. After their Sunday afternoon visit at his Aunt Angèle's house, he would cling to *Maman*, crying and raging, insisting that she take him with her to live in the hotel, while she protested the impossibility of such an arrangement. This time, however, Bunny did what Hélène would or could never do:

> I gave in, even though I knew I was making a mistake. He was irresistible, and I still loved him, and I couldn't bear to see him like that, and I thought, what if he really did kill himself? The hardest thing I've ever had to do was to go down and tell my friend who was still waiting in the taxi that I wasn't going to the Virgin Islands with him after all. The reconciliation was beautiful—they always were. We left New York City and rented a little house in Amityville, Long Island, and I became pregnant. Even though Jean had agreed to the idea of children, when I told him, he was furious, and although he finally accepted the idea, he was never really happy about it. It was a difficult pregnancy and a difficult birth, and it was not long afterwards, or maybe even during the time I was pregnant, that he got involved with someone else and decided he wanted out. It was for good, this time. The breakup was terribly painful. My feelings about him weren't resolved for a long, long time. I said to myself later, how naive I was ... because I had really believed that if I left him, it would kill him.

In these unhappy circumstances, in November 1950, Laure van Heijenoort was born. But while her mother described Jean's reaction to the event as less than enthusiastic, van Heijenoort himself was to tell his closest friends that the two most important women in his life were his mother and his daughter. It may be odd to hear of him expressing his feelings in such conventional terms, considering his strange and unconventional ways of demonstrating or concealing affection; nevertheless, it was true. These two women did have a powerful emotional hold upon him that he embraced. Still, as with Jeannot, he was an uncomfortable father who did not quite know how to play the part. Removed, but always somewhere on the periphery, he kept watch, without being directly involved.

*Father and daughter (Laure). New York, 1956.*

Laure van Heijenoort's reflections as an adult have corroborated this detachment. While she sensed that her father loved her deeply, she never felt she knew him, never felt he understood her, and except for a few rare occasions, never felt close to him until the last troubled months of his life. To her he was very shy, very closed, and almost anti-social: "He didn't communicate much to me or to my friends, especially to my boyfriends—he wouldn't talk to them at all." Laure was sixteen years younger than her stepbrother, but her description of her father was much like Jeannot's, and like him, what she knew best and impressed her most as a child was her father's skill at building things.

* * *

With Loretta "Bunny" Guyer, van Heijenoort established what would become a pattern: on the verge of breaking up a marriage or a long-term relationship, he would maintain a certain level of commitment that, at random moments, made it seem as if all he desired was to surround himself with the standard accoutrements of domestic stability. Thus at the same time Van and Bunny were about to divorce, he was—with his "own two hands"—building a house in Amagansett, in the township of East Hampton on Long Island.

The idea for the project grew out of an invitation from the art critic Harold Rosenberg and his wife May, who had recently moved into a house in that area. Following the example that André Breton and the French expatriates had established during the war, van Heijenoort and his friends continued to make a regular thing of going out to Long Island on weekends and holidays. Especially in the summer, there were frequent excursions and house parties. A few of van Heijenoort's friends from the *Partisan Review* crowd and the Eighth Street Club rented houses together, and eventually some built places of their own in the area known as The Springs. Among these were the Rosenbergs, the artists Jackson Pollock and Lee Krasner, Willem de Kooning and Franz Kline, and Benjamin Gollay, Van's good friend and later his lawyer.

In 1948, after a weekend with the Rosenbergs, the van Heijenoorts, on impulse, decided that having their own place somewhere on that part of Long Island would be a good idea. From impulse to action, they wasted no time. Within weeks, they found a thirteen-acre parcel near their friends and bought it for the grand sum of one thousand dollars, borrowed from Bunny's parents.

In East Hampton, the story of the house that Jean built, almost single-handedly, became one more chapter in the history of his mysteriousness. His friends—people in academic or intellectual circles—were very surprised to learn that he knew how to do anything practical because his focus was always on the cerebral and the abstract. Everyone asked, "But where did you learn how to build a house?" Van had two answers, usually given one after the other in no particular order, and both only partly tongue in cheek: One was, *"Mais il n'y a rien qu'un vieux bolshevik ne puisse faire"* (But there is nothing an old Bolshevik can't do), which was the answer he also used when he was asked how he knew how to fix cars or build a well and so forth. The second explanation was: "Well, you just take a book from the Forty-second Street Library on how to build a house. It's easy."

(or, as Stefan Bauer-Mengelberg told it, in Van's accent: "...'ow to build a 'ouse"). Both statements were punctuated with a beatific smile and a remark indicating that here was one more justification for his life-long love affair with American libraries in general and the Forty-second Street library in particular.

In truth, besides Bolshevism and book-learning, van Heijenoort had some actual practical grounds for confidence. "As a boy, I was always building something or fixing my bicycle or putting something together," he confessed later, and it seems to have always been a "something" of great importance to him. One of his half-sister Paulette's strongest childhood memories of her brother during the short time they lived together was his admonishment to never, but absolutely never, touch anything in his room—if, that is, she was granted permission to enter at all. And until the end of his life, Van was a tinkerer and an avid Radio Shack shopper.

Nothing gave Van more satisfaction than the achievement of building his small house. He was openly proud of having done it and never tired of describing the process in precise detail, not unlike an exposition of a mathematical proof:

> The main part of the house was thirty feet by twelve feet and then there was an addition for a kitchen and a bathroom that was eight feet by sixteen feet—eight by eight for the kitchen and eight by eight for the bathroom. It was all in a line. The kitchen was perfectly all right: I had a big sink and a frigidaire and a stove and I did the bathroom too, the shower and the sink and the hot water heater. I had a kind of butane lamp to do the soldering. It was quite simple. The foundation was a series of cedar posts sunk six feet into the ground. I dug the holes myself, by hand, with a special tool. You just put in your poles and you put stones around them and after a while they are just solidly there. The thing is, they have to be very well aligned because you have to put your big beams on them, and then you put your floor and the walls and the rafters, and then you put on the roof. After I put on the roof, I placed a bunch of flowers and leaves on top to celebrate that that much of the house was finished and closed in. The roof was tar and gravel and special tar paper with a lot of overlap. It was perfectly tight, never leaked. I never had one drop of water

*Jean van Heijenoort on his land. East Hampton, ca 1955.*

in the house. In 1954, there was a huge hurricane and another one in 1956 and many houses around me were damaged, but not mine. Of course, I was protected by the trees because I had cut as few trees as possible.

The only power tool Van used was a six-inch skilsaw. In France when he had made something, it had always been with a hand saw, because that was all there was. "But I could never have built a house sawing by hand. Never." Van said later. "I had to have a power saw, but that was the only electric tool I used. Otherwise, I had a hammer, a chisel, and a hand drill." Working in that fashion, it took the old bolshevik three summers to finish the house completely, but by the end of the first summer he was able to live in it in what he called a "rough way," equivalent to camping.

It is entirely possible that van Heijenoort thought building a house would revitalize his shaky marriage and, in fact, for a short period he and Bunny worked together on it, but by

*The house that Jean built in East Hampton.*

the time it was finished their marriage was too. Laure, as a toddler, came to the house with one or the other of her parents but never with both. At first Bunny and Van took turns, but eventually it became exclusively Van's domain, his retreat from New York and, later, Cambridge. If there was any place that he thought of as home, this little house in The Springs was it, a felt reminder of the landscape of his native territory in northern France. It was a base he would touch for more than twenty-five years, and even afterwards he talked about it with a pride of ownership quite rare for him: "I used it all the time. It was the only way for me to survive New York. On Thursday or Friday, I would take my car or the train, and arriving there, you could breathe fresh air."

The "Heast 'ampton" (as he pronounced it) and Springs environment provided Van with more than a release from the stifling city and an elevation of spirit that came with country living in a neighborly community. Nowhere did the many facets of van Heijenoort's professional and personal life come to light as they did in his small house, on his land. There he was the scholar-professor-teacher, the do-it yourself builder, the gracious host, the lobster cook, the sometime-father, the flirt, the romantic lover, the sometime-husband, and even, the loving son, for in time *Maman* Hélène would come to visit. Taking her first airplane flight, Hélène, now in her eighties, crossed the Atlantic and had the sweet pleasure of seeing her son Jean in his element—the scholar, surrounded by people who obviously respected and admired him. *Maman* stayed in East Hampton and cooked elegant meals for Van, his colleagues, friends, and lovers. She also had many visits with her ex-daughter-in-law Bunny and her American grandchild, Laure. In her tactful, charming manner, Hélène accepted and loved them all, while *she* in turn was considered to be one of the natural wonders of the world by everyone who met her.

East Hampton turned out to be a very good place for Van to work, and it would be there, in the 1960s, that he would do most of the work on his famous "Source Book," with the assistance of his student-cum-colleague Stefan Bauer-Mengelberg as well as many others. It was also a romantic place to bring the succession of girlfriends that followed his divorce from Bunny. Something special seemed to happen to Van on Long Island. When these women spoke about the house in The Springs and the time they spent there, whatever they may have felt in the way of bitterness regarding other aspects of their relationship with Van seemed to drop away. It was a place to be quiet, to walk in the woods, to go to the beach, and to be part of a community of lively and interesting people. It was there that Van was most expansive, in visiting others or inviting friends.

* * *

During their long years of friendship, Stefan Bauer-Mengelberg, Van's student and colleague, became a skilled imitator of Van's accent, gestures, and facial expressions. When Stefan recounted an episode that involved Van, it was a bravura performance with quotes done in Van's voice. He would even tell these stories in Van's presence, while the latter blushed and laughed and to all appearances enjoyed the show as much as anyone. Here is one of Stefan's twice-told tales:

> Van and I often took the Long Island Railroad from New York on a Friday afternoon. One time Van and I and two other friends were sitting in the club car which in those days had nice old fashioned chairs. Somewhere around Mineola, Van spied a nice-looking blond girl sitting across from us, so he said, "Excuse me," and got up and went over to talk to her. All the way from Mineola to Bridgehampton, which means almost three hours, he sat with full concentration and rapt attention. Finally, she got off at Bridgehampton and he came back to us and sat down with a sigh, saying, "You know, zat girl do not 'ave a brain in 'er 'ead."

As a footnote to this story, Stefan added (not in Van's presence) "Only Van could do that. He seemed to suspend all his critical faculties when he was in the presence of a woman to whom he was attracted. And I think he had a positive tropism for kooky women ... nervous, high-strung types."

Some of Van's women were intellectual, some were not; most were at least twenty years younger than he. His "type," as *he* would describe it, was slim, dark, and boyish, and bohemian rather than "kooky."

The first woman van Heijenoort brought to Long Island after his separation from Bunny was Anne B., a strikingly beautiful undergraduate from New Jersey, who enrolled in his math course in her freshman year in order to satisfy one of the requirements for her liberal arts major. She fit his general "type" requirements so very well that he immediately fell in love with her. Anne, eighteen, and, by her own characterization, "unformed," had her preferences too: "Father-types and teacher-types," she said later, "men who were very intellectual, who I felt I would learn from." Can there be any wonder that she, too, fell "madly in love" with the elegantly French, soft-spoken, professor of mathematics? Because she had always been attracted to older men, not only was the twenty years' difference in their age not a problem for her, it was exactly what she wanted. Her parents, on the other hand, were distressed because of Van's age, his marital status and the fact that he had children. But in spite of their strenuous objection and their attempt to persuade Van not to marry their daughter, Van and Anne were married in 1952, not long after his divorce from Bunny.

The ceremony took place in City Hall with the apprehensive parents in attendance, after which they all went to a French restaurant where the eighteen-year-old Anne ordered *tête de veau*, in order to impress everyone with her sophistication. "When it came, I almost fainted," she later confessed. "I didn't know what to do because I didn't want to say that I hadn't known what I was ordering, but I knew I couldn't eat it."

Otherwise, Anne thought, the marriage began well:

> We rented this third floor cold water flat on West Third Street. I thought our apartment was wonderful, and the other occupants of the building were also wonderful. I felt very bohemian. On the bottom floor there was a man who was a Reichian; he was making orgone boxes and torturing mice. [The orgone box, designed by Wilhelm Reich, was made of wood, tin, and other materials such as compressed vegetable matter. It was alleged to restore "orgone energy" to those sitting in it, thereby aiding in the cure of impotence, cancer and the common cold.] On the next floor, there was a man named Herbert who was a homosexual and incredibly nice. I met my best friend through him.

According to Anne, "the wonderful beginning lasted about three days and then Jean said, 'I can't stay,' and just went off somewhere—I had no idea where. I stayed on in the apartment and after a while, he came back to me."

Three days was something of a joke; their marriage lasted four years but, in fact, for much of that time they did not live together. Rather, the pattern, established almost immediately, was like the one late in Van's marriage to Bunny: they broke up and got together only to separate again, reconcile, and repeat the cycle.

After their first reconciliation, the couple went to East Hampton for a summer. Anne remembered that as a romantic time when they were still very much in love:

> It was a Pygmalion situation; he wanted to form me, to teach me things; he wanted me to learn, he wanted me to blossom. He really loved teaching, and what he liked about me was that I was very young, very naive, and willing to do almost anything he suggested. He assumed that I had certain qualities which I

> may or may not have had. In any case, he had an enormous effect upon me. I did not know anything about literature or painting, and I was eager to learn and eager to please him. My whole life changed. He introduced me to all his friends in East Hampton who were very intellectual—of course he knew a lot of the writers and painters and all the *Partisan Review* people. Half the time I didn't know what they were talking about, and I was very intimidated and shy. He also taught me a lot about food [presumably about *tête de veau*], about how to cook, because I didn't know anything about that, either. He even encouraged me to start painting. I'm not very artistic, but I loved doing it. He was never critical; he said he liked the things I did.

The environment of East Hampton was conducive to happy times. It was something like camping out, particularly the first year Anne was there, when Jean's house was finished but had no interior toilet. Instead there was an outhouse that he kept moving to different places. By the second year, however, he put in the kitchen and bath; then their friends, Charles Boultenhouse and the poet Parker Tyler, who lived nearby in a shack without a toilet or electricity, would come over to use the facilities and stay on for the evening. They had poetry readings by Tyler and skits put on by Anne and Charles; there were invitations to parties at Willem de Kooning's and Harold Rosenberg's houses, and beach parties where, as Anne later said, "half the people were bathing naked and the other half were having these very intellectual discussions." While Anne was in awe of everyone—the de Koonings, the Motherwells and the Pollocks—she also noticed that the artists were in awe of her husband. He was odd, difficult to approach, not gregarious, yet he seemed to want to be part of their group. They talked *about* him more than to him, but he was accorded great respect. Although no one had any basis to judge, everyone assumed he was brilliant because he was a mathematician-logician-philosopher; and also his long and close association with Trotsky never failed to make him intriguing.

The two long summers Anne and Jean spent on Long Island were the good times. The rest was awful. What had been "beautiful and pure" fell apart as soon as anything disturbing arose. When the honeymoon was over, it was as difficult for him to sustain a daily give-and-take domestic arrangement with Anne as with any woman. He distanced himself, first emotionally and then physically, by going away for an indefinite time. Anne B. found him incapable of compromise. When he wanted something or someone, it had to be his way. "He was never willing to change," she said later. "You couldn't get to him. He was like a wall. What you thought, didn't matter." She judged him cold, selfish and even cruel.

Nothing seemed to stimulate Van's interest so much as the discovery that another man was interested in his wife or girlfriend. Referring to her husband's refusal to let go at the moment when she was ready, Anne noted, "He would leave me and then come back, supposedly 'just to say hello,' as soon as he found out I was seeing someone else." As with Bunny, there were extravagantly jealous scenes and dramatic pleas for reconciliation, and like Bunny, Anne, too, would take him back. "I stayed involved with him for many years when I shouldn't have." she said later. "When I think back, there was something a bit fakey about him. It's hard to know how much he was really feeling and how much he talked himself into, because he was *very* romantic."

Romantic first, detached and distant later, van Heijenoort never elaborated on the reasons for his disaffection from Anne or any other woman. He never acknowledged that ordinary, long-term domestic arrangements were not to his taste, although clearly they were not. At most, he would imply by a tilt of the head and a roll of the eyes that he was a man more sinned against than sinning. Still, the best explanation may be his credo: "If there are no strong feelings, there is no point. I am a man who believes in strong feelings."

The time did finally come when Anne and Van were both ready to dissolve their marriage. He had another girlfriend, and Anne had met the man who would become her second husband. In 1956, their marriage was annulled. Each attributed the decision for annulment to the other: Van said, "She wanted it that way," whereas Anne said, "I'm sure *he* came up with the idea because I wasn't planning to do anything. Maybe he chose it to save money." She did nevertheless acknowledge that she preferred an annulment "because it made it seem like we had never been married."

There were several objective advantages to having an annulment as opposed to a divorce. Prior to 1970, a no-fault divorce could not be obtained in New York; one had to prove adultery. Benjamin Gollay, Van's friend and lawyer, had handled his divorce from Bunny; he was also trying to help Van with his application to become a United States citizen—and to keep him from being deported when that matter was threatening. Gollay thought another divorce would add fuel to the deportation advocates' argument and certainly would not help on the citizenship front. It was he who suggested the annulment as a wiser course, and Anne and Van agreed to stipulate to whatever the law required.

Although she was glad to be out of the marriage and had no quarrel with the method of dissolution at the time, Anne retained bitter feelings about it. As part of the agreement, she signed a document stating that she would never make a claim on the house in Springs. Many years later, the issue still rankled her:

> As if I would have done that ... but Van was always petty about money. He was known for not being generous. He never spent any money and people would imply that he sponged a bit. His friends even used to kid him about it. I didn't care because I had always worked, the whole time I was going to school. He could have said, Look, I don't want to be with you, but let me help you. At the beginning, I asked, and he did give me something but otherwise, never.

Anne B.'s statement that van Heijenoort was stingy provoked sharp disagreement and some evidence to the contrary from his son and daughter. According to Jeannot, his father sent him money after Gaby's death, while he was still in school and during his military service. He also contributed to *Maman* Hélène's support. Laure van Heijenoort maintained that her father was never anything but generous as far as she was concerned. In her view, his personal spending habits were a reflection of his values. He spent money on the things he thought were important: books, travel, healthful food, and medical care. Both Jeannot and Laure emphasized the fact that their father grew up in hard times and in a particularly difficult family situation.

But several of van Heijenoort's close and affectionate friends agreed with Anne B. He was a much more frequent guest than host, and in a restaurant he never leapt for the bill. Stefan Bauer-Mengelberg, who, in general, made it a point of honor to pay the check when he and Van ate out together, had a story to corroborate Anne's point:

> Van had been away in Europe for several weeks, and I had taught at least four classes for him and had even corrected papers. When he came back, he proposed that we meet for dinner at Monte's, which we did. I took this to be his way of thanking me, so this time when the bill came, I hesitated, and then Van said, "Well ... let us divide it."

* * *

In the mid 1970s van Heijenoort sold his house in The Springs. Because he was teaching at Brandeis and, after 1969, spending several months a year in Mexico, he rarely went to Long Island. For a while, he rented his house to others, but then it seemed to him there was no point in keeping it. Afterwards his friends said they were not aware of his deep attachment to what he had relinquished. Stefan, for example, was surprised to learn, ten years later, that Van had expressed regrets about the sale, and that he had said his house on Long Island was the only place he ever thought of as home:

> He said nothing to *me* which indicated that he had any feelings at all about it. The transaction was handled by a real estate agent, Van was absent, and he never showed any interest in meeting the young couple who bought it.

The new owners, Joseph Stillman and Minna Kotkin, had no idea whose house they had bought until after the sale. They were taken to the property by a real estate agent who said, "I have something very odd in my listings. Would you like to look at it?" The Stillmans later recounted their first impressions:

> When we drove up, there were three sort of middle-aged Hispanic women sitting at the picnic table outside, wearing black tights and leotards, looking very severe. It was a dark, hot, muggy day and it felt something like coming upon Macbeth's witches. We thought the whole thing—the house, the atmosphere, the land—was wonderful. At that point we would have signed anything the agent handed us.

In spite of his very positive attitude towards the property, Joseph Stillman's description of the interior of the house was markedly different from van Heijenoort's cheerful focus on the marvels of the kitchen and bathroom:

> It was like a dark little cave inside. In one corner there was a platform for a double bed and then there were four or five very large tables; you had to thread your way in and around them. My sense of it was that it was a place someone came to *do* something, not just to be there.

The rest of the house got mixed reviews from the Stillmans. The framing was excellent; after thirty years everything was still at perfect right angles. When Stillman went to put in a new floor, he didn't have to trim any boards to make them fit. However, other details of the construction, particularly the plumbing and the electricity, were considerably less exact and the new tenants had some surprises in store for them. Stillman said,

> The house was plumbed from the well with a piece of garden hose, so you could have water from only one faucet at a time. After we had been in the house for a few months, the toilet stopped flushing. I called a plumber and told him I thought we needed a new septic tank. He went out back to have a look, came back and said, "There is no septic tank." There was just a pipe, running out into the ground about twenty feet from the house. It *had* seemed kind of squishy to me out there but nothing too repulsive. I guess everything had been draining down into the soil for years, but finally it began to back up.
>
> Another problem was the electricity. He had wired the whole place in lampcord and underneath the house there were skeins

> of frayed lampcord hanging down into leaves. It was absolutely terrifying. When I went to redo the floor in the living room I found that the kerosene heater was leaking. The whole house seemed to be an experiment in spontaneous combustion.

This carelessness or obliviousness to hazard was a rare step out of character for van Heijenoort. But otherwise, just as he said, the roof never leaked and the cedar foundation posts sunk into the ground held firm and true. When the Stillmans decided to enlarge the little house and make it double the size, they built their addition in a style so much in keeping with the original that from the outside it is difficult to discern which is the new and which the old.

* * *

Whether or not Van felt pangs of regret after he sold his house, there is no question that his place had meant a great deal to him. Later in life, he continued to talk about the Long Island scene: the flora, the fauna, the marine life, the lobsters he cooked, and most of all, his friends and neighbors, particularly the painters, whose work he admired before they became famous. He had felt a special affinity with the silent Jackson Pollock, but found his drinking bouts and self-destructive behavior incomprehensible. Willem de Kooning, whose work Van liked at least as much as Pollock's, made him much less nervous.

Van also continued to dream about his house and in his subconscious world, one can find his satisfaction, his nostalgia, and his sorrow:

> I dream about my house in Springs. I have improved it. I have made it larger. I have added another room and everything in it is just as it should be. I stand in the old part and I look into the new part, which I myself have built. But I don't go into it. I am not supposed to. I am not allowed to use it.

Chapter Thirteen

# A Passion for Exactitude

1957–1977

The *Source Book* was a lifetime of labor, I thought. All you had to do was look at it casually to see the overwhelming amount of work it must have taken. In fact, it made me tired just to look at it—to feel the effort that went into it, the care, the years of scholarship and reflection on these matters.

JUDSON WEBB, *1986*

AMONG A CERTAIN group of scholars not ordinarily given to hyperbole, the 1957 Summer Institute of the Association for Symbolic Logic held at Cornell University, in Ithaca, New York, is spoken about as a watershed event, a once in a lifetime, impossible to duplicate gathering. It was, and remains, unique partly because it took place in an era just before scientists began to spend their summers trekking around the world from one meeting or symposium to the next. That summer, for four weeks, an esoteric coterie of logicians, who had been happily working away in their own little enclaves, walked and talked together on the beautiful Cayuga Heights of the Cornell campus and presented each other with new ideas, methods, and intuitions, and new ways to look at old problems.

The resulting cross-fertilization was to bear fruit for years to come, and the kind of excited interchange that was begun there would be taken for granted at future major conferences. The proceedings of the meeting were collected and bound into a heavy hardback volume of 427 pages, which would be proudly carried about and studied by many a graduate student and postdoc too young to have been present at the time. If ever a logic meeting evoked nostalgia, Cornell did.

It was a small gathering: eighty-five registered participants: stars, superstars, rising stars, falling stars, and just normally smart logicians. Alonzo Church, Haskell Curry, Stephen Kleene, Willard Van Orman Quine, Abraham Robinson, Barkley Rosser, and Alfred Tarski were among the illustrious names. Only the reclusive Kurt Gödel was noticeably absent. Fifty-five scholars came to present one or more papers and the rest only to listen, but for all, there was plenty of time for informal exchange, and time to get a feeling for the novel work and accomplishments of others.

Jean van Heijenoort came as a listener–participant. He was forty-five, but, because of his late start, much less established in the field than many of his colleagues who were fifteen and twenty years younger. However, he did know the basic literature and had already written many reviews of articles on logic and the philosophy of mathematics. In referring to the meeting in later years he marked it as a seminal event in his professional life. At Cornell, besides the important senior logicians of the day, he met many of the people he would work with or be strongly influenced by in the decades to follow: Burton Dreben, of Harvard, who brought the work of Jacques Herbrand to his attention; Solomon Feferman, of Stanford, with whom he would collaborate twenty-five years later, as an editor of the collected works of Kurt Gödel; Charles Parsons and Hao Wang, who were to assist him with *From Frege to Gödel*, the work that was to bring him great acclaim; and Georg Kreisel, then of Reading University in England, whom Van would later cite as the man who influenced him most ... after Trotsky.

The juxtaposition of Trotsky and Kreisel may seem surprising to those who know of Kreisel's elitist attitudes, yet there were similarities between the two men. Kreisel was concerned with logic and theorems, not politics and political theory; his "mission" was entirely different from Trotsky's in nature and scope, but his forcefulness and charisma were reminiscent of the Old Man's. A persuasive leader, a man with decided views, Kreisel liked to expound general ideas; he had strong opinions about whose work in logic was worthy, and stronger ones about whose was not. An impatient guru, he nonetheless attracted and encouraged followers.

Although he thought of Kreisel as a great influence, van Heijenoort did not actively engage himself in Kreisel's programs. He did, however, take Kreisel's ideas and explorations seriously. Kreisel liked to talk and Van liked to listen; it made their relationship "very congenial," as Kreisel has said. And, with a certain apologetic defensiveness, Van professed "very warm feelings for Kreisel." His manner implied that an apology was expected because of his friend's reputation for being difficult. Like Trotsky, Kreisel had a rapier wit that the restrained Van appreciated. Perhaps he also took vicarious pleasure in his colleague's occasionally outrageous expressions.

Despite very different personalities, there were many affinities between the two men. They were Europeans, both educated in boarding schools, both separated from their parents at an early age; Kreisel, an Austrian, had been sent to England during World War II. The important specifics of their experiences were not at all the same, but each had an intuitive understanding and sympathy for the problems of the other. Both men were extremely sensitive to noise and had trouble sleeping, and they enjoyed discussing their sleep and general health problems in great detail. They corresponded voluminously, writing about personal and professional matters, and saw each other regularly in Europe and in the United States.

Stimulated not only by Kreisel, but also by the heady atmosphere of Cornell, by what he had heard and learned—not, as was his custom in the past, from books and papers but from the sources' mouths—van Heijenoort returned to New York University eager to do something of his own in logic. Unfortunately, at NYU there was no one with whom he could share his enthusiasms, no one whose interests paralleled his, no one to test his ideas about what he might contribute to the field.

For the purposes of instruction, the mathematics faculty at New York University was separated into the graduate school (later to become the Courant Institute) and the undergraduate school, which was then headed by Morris Kline, a man whose attitude towards logic was less than enthusiastic. Van Heijenoort's appointment was in the undergraduate college, and he had almost no contact with the graduate school faculty. His major responsibility was to teach the classic, garden-variety, introductory math courses such as analytic geometry and calculus. Although he was more than competent in that regard, his heart was not in it for long. In Stefan Bauer-Mengelberg's opinion, van Heijenoort was a good teacher for bright students but impossible for bad ones because he had no patience—and perhaps no understanding—for anyone who might have difficulty with mathematics. "When I asked him how his regular classes were going, Van would answer with a shrug and a sigh, 'Oh, zey do not care, and I do not care.'"

Nevertheless, Van was able to introduce one new subject into the undergraduate curriculum which did interest him very much. Every other year, beginning in 1949, he had taught an undergraduate course in the foundations of mathematics. Bauer-Mengelberg wore his participation in that class as a badge of honor. He considered himself to be one of the best students because as a philosophy major and a mathematics minor he was especially interested in the subject. Isaac Levi, later professor of philosophy at Columbia University, also took the course and was equally involved in the material.

Levi felt there was something deeply personal in van Heijenoort's interest in logic: "He had a yearning for objectivity and he thought that logic was a place that was ideology-free." The observation is particularly interesting because Levi and Bauer-Mengelberg were members of a small circle of students who defined themselves as members of the anti-Stalinist left. To them, van Heijenoort was a hero, not because he ever talked to them about anything political but simply because they had heard about his past connection with Trotsky. That was enough to stimulate their imagination and to give them the feeling that they intuitively understood what was seething beneath their professor's controlled countenance.

Dissatisfied and disheartened though he was at the general prospects for doing logic at NYU, van Heijenoort did not try to improve his situation by looking for a position elsewhere. As long as his residency status vis-à-vis the US Immigration Service was uncertain, he schooled himself in caution. He decided that this was a time to stay put, to appear to be the stablest of men. It was not an easy assignment for the nomadic van Heijenoort who, no matter what he said or tried to do, was ever inclined to change his residence, his wife, or his lover. But finally, in 1958, after eighteen years in the United States, Jean van Heijenoort was granted citizenship. At last it was decided by the "powers that be" that the ex-Comrade Van, now Professor van Heijenoort, was an asset rather than a threat to the security of the United States.

Not surprisingly, the first and major item on the new citizen's agenda was to leave the United States—temporarily, but now safe in the knowledge that he could return. His desire to go home again to Creil, to see his mother, his son, his half-sister, his cousins, and his old friends could, at last, be satisfied. Before the 1958 spring semester was over, he had made his plans.

Although van Heijenoort frequently masked his feelings, he did not underplay the drama of his first homecoming. For all his ambivalence and all his feelings of alienation from early childhood onward, France, Creil, and Paris were still home. Twenty-five years later, with considerable emotion, he recalled the day:

> It was in the middle of September [1958], a hot day. I had flown to Brussels and traveled by train from Brussels to Paris and from Paris to Creil. I was terribly excited, everything was so strange, yet so familiar. The roofs of the houses were what struck me first. They didn't look American, not at all. They were European roofs.
>
> There were delays, so I didn't arrive at my mother's house until nine at night, much later than I had planned. I had not given any precise time or date for my arrival, but I had written that I would be coming soon. It was already dark and my mother was getting ready for bed. I pushed the bell. She came out and looked at me, and finally, after twenty-two years, we embraced. We both cried. Then we went inside and she fed me some Camembert.

In the days that followed, van Heijenoort became reaquainted with the son he had not seen for ten years. Jeannot was now twenty-three, taller than his father—a man just one year younger than Van had been when he left France for Mexico at the end of 1936. He saw his half-sister, Paulette, who was sixteen when he left and was now a wife and mother; and he visited the cousins with whom he had lived as a child, who were still in the very same house on the edge of town. Paulette was surprised at how little her brother had changed, except for his accent which she thought was American. (Van's colleagues at NYU would have laughed to hear her say it.) "Physically, he was the same," Paulette reported later. "It was *him*. We were all so moved to have him with us."

Casting a shadow on the warm family feelings surrounding Van's homecoming was the poignant death of Gabrielle Brausch just six months earlier. Gaby was only forty-eight when she died of heart failure. She had been in poor health since her wartime imprisonment and had never fully regained her strength. Van Heijenoort had admired her frankness and her pluck, and always

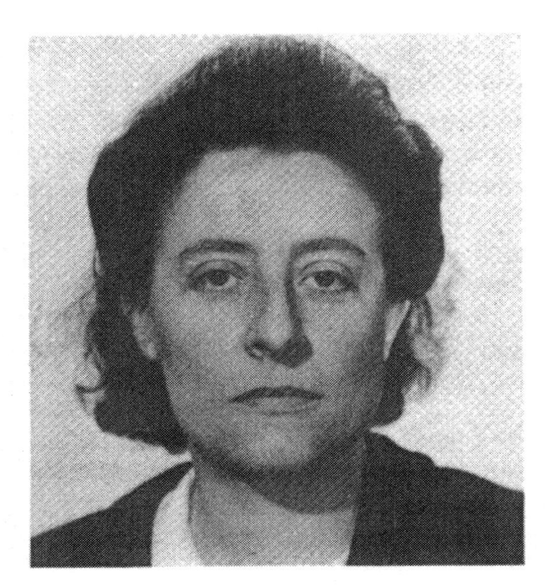

*Gabrielle Brausch. ca 1950.*

described her as a very courageous woman. Without his having said so explicitly, one sensed that he was left with a residue of guilt about his separation from Gaby twenty years earlier and the subsequent trauma in her life.

The "American" Jean made his rounds of the family and was introduced to Paulette's husband, Manuel Corpion, and to their children. He had a reunion with his closest childhood friend, Roland, the only one of the *quatre Creilleois* who continued to live in Creil. The closeness was still there, along with the old habit of silence. Roland was heard complaining: "The person who can get Jean to reveal what he is thinking has not yet been born." So Jean was indeed the same, only more so. As Paulette said later:

> For us he was always some kind of a star, but now more than ever we understood very well that he had evolved, that he was in another world intellectually, and financially too, but except for subjects concerning the family, there was little he wanted to tell us about his voyages and his life during those twenty-two years he was away.

True to form, van Heijenoort maintained his customary guard and spoke in evasive generalities about his own extraordinary political and personal experiences, never explaining the reasons for his rupture with Bunny or the complexities of his separation from Gabrielle Brausch. On the other hand, *he* wanted full and precise descriptions of what had transpired in Creil during the years of his absence, including all the details about what life had been like during the Second World War.

In the wake of the excitement of being home again as Jean rather than Van the exile, he embarked upon a series of excursions to the haunts and houses of his youth. Reacquaintance with the physical contours of the old familiar territory and the changes that had been wrought was as important to him as the human contact. He visited the site of the grim boarding school in Clermont de l'Oise only to discover that the school had been bombed during the war and was in ruins. He went to the house in Saint-Quentin where he had lived only during vacations, and to the cemeteries and battlegrounds of the First World War which had had such a profound effect upon him in his youth. "You could feel how very attached he was, to all the old places, even after all that time," his sister said. "He had to find them all."

From that first visit in 1958 until his death, in 1986, van Heijenoort was to return to France twice a year, usually for part of the summer and often in the spring, always visiting his mother, his son, and the various members of his immediate family and always spending a certain amount of time in the *Quartier Latin* of Paris. In the years that followed he brought at least two women, with whom he was romantically involved, to be introduced to the family. The charming *Maman* embraced them all, setting the tone for the rest of the family. Although others may have raised eyebrows, she accepted her son's choices and way of life without question or judgment and, almost always, with enthusiasm. She was relieved that he had moved away from politics and returned to his early passion for mathematics and philosophy. Regarding his love affairs, she said, "Et bien, les hommes." ("Well ... men.") As for his children, she doted upon them and they adored her. Since his return to

France, Jeannot had spent a great deal of time in her house, and now Laure van Heijenoort, too, became a frequent visitor. After 1958, sometimes with her father and sometimes alone, she went to stay with her grandmother in Creil. Although as a child and teenager Laure's relations with her father were not easy, she loved her French grandmother and her half-brother Jeannot, who still spoke English and acted as her interpreter and protector.

One of the women Van introduced to his family in Creil, in the early 1960s, was Beverly Woodward, his girlfriend, companion, and, after their romance ended, his good friend. They met when she was a student, a philosophy major who wanted to learn more mathematics. It was natural for someone to recommend van Heijenoort to her, and soon she was sitting in on several of his math courses. Like others, before and after, Beverly was smitten by her professor, and he found her very attractive. She was young, dark-haired, slim, and intense—once again, his "type." She also knew German and French, a great virtue in van Heijenoort's eyes, and he enlisted her as a translator for the *Source Book in Mathematical Logic* on which he was working.

By then van Heijenoort was forty-nine; she was twenty-eight, not as young or naive as Anne B., but the dynamics of their relationship were similar. She was the bright student, he was the mentor-professor. They lived together, on and off, in his apartment in Washington Square, and they traveled to Europe together in the summers of 1961 and 1962. Beverly observed that Van was much more in his element in France, more comfortable. People knew who he was; he didn't have to explain himself. She also noticed how eager Van was to spend time with his mother and how exceptionally loving and tender he was with her.

Besides philosophy and mathematics, Beverly studied music. She was a gifted pianist and, during the time she lived with Van, began to consider a career as a performing artist. This meant that she needed to spend a great deal of time practicing, and the problem of when and where she could do so without disturbing Van, who was extremely sensitive to sound, became a constant source of conflict. Also, according to Beverly, Van became irrationally jealous of her piano teacher, and there were terrible scenes about that. Beverly and Van came into conflict over politics, too. Beverly was active in the peace movement and would later become the coordinator of an institute for the study of non-violent alternatives. She was an optimist who believed in the possibility of change, whereas van Heijenoort was a pessimist and skeptical about the long-term prospects for peace. After some particularly horrendous world news, he would say, "Well Beverly, it does not look too good for nonviolence these days." Beverly, committed to her peace work with the kind of intensity Van had once had for his Marxist ideals, often got extremely angry at him over political issues. By the end of 1962, their romance foundered, ostensibly because they could not find a place to live that satisfied all their requirements. They did however remain close and affectionate friends, and were in frequent touch with one another.

* * *

The carrier of a passport issued by the United States of America has a great deal of freedom to move about the globe unhindered. The newly minted citizen "John" van Heijenoort, for that was the name he had used on his citizenship application and his offical name at NYU, took full advantage of this right. He went regularly to France and Mexico. He also went to India for several months, to Turkey to revisit the scene of his political past, to the Netherlands to seek out his father's relatives (the van Heijenoort family he had never met). In Holland, van Heijenoort also went to the Amsterdam Institute where one of the major collections of Trotsky's papers is housed. This was a business visit and had little to do with nostalgia.

In the mid 1950s, through the instigation of Douglas Bryant, the Harvard librarian, van Heijenoort had agreed to help with the acquisition of documents for the Trotsky Archive at Harvard University Library. Before his death, Trotsky had sold his papers to Harvard, but until Bryant was hired in 1952, nothing had been done to make these papers available to scholars. Bryant's first specific assignment was to get the post-1936 Trotsky archival material out of Mexico, and in general to see to the organization of all the Trotsky papers. Quite naturally Bryant contacted Jean van Heijenoort, then living in New York, and persuaded him to accept the position of "specialist in acquisitions." Although the former secretary had renounced Marxism and Trotskyism, he could not refuse to participate in the gathering of what he considered to be important historical material. He was paid little or nothing for his years of work on behalf of the Trotsky Archive—at times it came very close to being a second full-time career—but he did receive attractive perquisites such as a small office in Pusey Library at Harvard, that was his to use in perpetuity, and access to all the facilities at Harvard.

After 1958, on one sort of business or another, mathematical, philosophical, archival, or personal, he was continually on the move. Sometimes his travel expenses were paid by his university or the sponsoring organization of a conference, sometimes by the Harvard Library fund for the Trotsky Archive, and occasionally, he paid his own way. As nomads make their seasonal rounds, he did his regular tour of the places important to him. It became an essential part of his life. While he spoke of settling in one place, of buying a house or an apartment in Paris, of living with one woman, he wandered. Unencumbered by possessions, once his teaching commitments for any given year were satisfied, he could and did pick up and go.

It took time before the former Comrade Van reestablished contact with his old Marxist "family" in France, except for an early visit with Jeanne Martin, Liova Sedov's companion. He went to Jeanne for several reasons: first and foremost, he wanted to learn everything he could about Liova's mysterious death at the Russian hospital. Also, in Van's opinion, Jeanne had crucial information about some very important papers that Trotsky had left in France; after Liova's death these documents had disappeared. As always, Van found it easier to solicit help from a woman; he counted on feminine understanding, and he was seldom disappointed. Finally, Jeanne was an old, old friend, one of the most intense links with the movement that had been his life. Sooner or later, that was another base he would have to touch.

* * *

For a few years in the early 1960s, while he was still a member of the mathematics department at NYU, van Heijenoort held a part-time appointment as visiting professor in the philosophy department of Columbia University. There he gave a first year graduate course in formal logic which attracted a wide range of interested students, from gifted undergraduates to faculty members such as Professor Sidney Morgenbesser. (Beverly Woodward sat in on that class as well.)

Morgenbesser liked and admired van Heijenoort and thought he was an excellent teacher: "He had good rapport with students in class and out. He was terribly concerned that they understand both the details and the significance of Gödel's proof." Others agreed, speaking of the benefits gained from van Heijenoort's drive and commitment to explain Gödel's theorem and to go through the entire proof step by step so that people would really understand it. Although van Heijenoort himself said that nothing in his life ever compared to the passion he had had for politics, he impressed all his students and colleagues with something very close to a passion for logic and for getting it right. One student recalled van Heijenoort's urging him not to read any other text on Gödel's famous incompleteness theorem before first listening to *his* (van Heijenoort's) exposition.

At Columbia, even more than at NYU, there was an aura of authenticity surrounding van Heijenoort, which was augmented by his reluctance to speak about political matters. All the loquacious young intellectuals arguing their fine points on Marxism, communism, socialism, Trotskyism, Leninism, Stalinism, revolution, dialectics, and so forth stood in awe of the man who had actually been there. And when, on rare occasion, he did say something, how they listened. His colleagues admired him for his modesty as well as his bravery; respecting his deep reserve they rarely asked him questions.

The pleasure of teaching the subjects that interested him most, at Columbia, made van Heijenoort ripe for a move away from NYU, where he had long felt underappreciated. His lasting distaste for living in New York City also contributed to his decision to leave. In 1965 he resigned from his position as associate professor of mathematics at NYU in favor of an appointment as professor of philosophy at Brandeis University in Waltham, Massachusetts, where he hoped to find a more hospitable atmosphere. Since symbolic logic straddles the fields of philosophy and mathematics, there was nothing particularly unusual about his change of department. At most universities, logic courses are listed in both mathematics and philosophy and, recently at some universities, in computer science departments. The students taking these courses may come from any discipline. Although historically the motivation for doing logic has come from philosophical thought, most of the significant work in the last fifty years has been heavily mathematical and most modern logicians are more comfortable in a mathematics department. In this regard, van Heijenoort was something of an exception. He had always had an inclination to do philosophy and had done his baccalaureate in both mathematics and philosophy. Since his interest in the latter never flagged, and he continued to read philosophy on his own, he was quite happy to be in a philosophy department as long as he had students interested in what he had to teach, and a few colleagues with whom he could exchange ideas and talk logic.

The change was propitious, not, as the situation developed, because of Brandeis University per se, but rather because of its location in the general Boston–Cambridge area, which as far as van Heijenoort was concerned was a cultural and intellectual paradise. He already had close colleagues at Harvard and, in the years that followed, would make new friends at Massachusetts Institute of Technology and Boston University. He attended seminars and weekly "Logic Lunches" at all of these places. His good friend and colleague Professor Burton Dreben frequently invited him to dinner with the Harvard Society of Fellows. Van Heijenoort enjoyed these evenings enormously, delighting in the academic banter and gossip. As Dreben overheard Van in conversation with others, he never failed to be surprised and impressed with the breadth of Van's knowledge; whether the topic was literature, politics, physics, current events, or ancient history, he always had something of interest to contribute to the discussion.

For the simpler pleasures, Van loved the intimacy of Harvard Square and would wax poetic over his current favorite ice-cream boutique. It was at once small town and high culture, and he liked being part of that community. Only the weather distressed him, especially in the winter when it was "cold, damp, and dark; worse than New York."

It is more than a little ironic that van Heijenoort left NYU because there was no program in logic, only to discover that at Brandeis, where logic *was* part of the curriculum, his symbolic mathematical brand was not. His closest colleague and ally was Jacques Cohen, a much younger man recently hired by the computer science department. Van Heijenoort and Cohen had many interests in common, ranging from logic's influence on computer science to the very practical consideration of how to use a portable computer to write a paper; Van's experience in the philosophy department was less felicitous.

Although van Heijenoort's Ph.D. students at Brandeis were in philosophy, most of his undergraduate students came from mathematics. That fact, per se, did not trouble him; but, since there was no program in symbolic logic in the philosophy *or* the mathematics department, even if these students became interested in mathematical logic (indeed, several were stimulated by their interaction with van Heijenoort) there was no way for them to get an advanced degree in that subject at Brandeis. The philosophy department was divided into two factions on this matter: those who defined logic in its more classical, restricted form, and those "moderns" who saw its significance in a wider framework of mathematical logic. Quite naturally, as a mathematician, van Heijenoort belonged to the latter group. For a while, he became engaged in a battle to broaden the curriculum to do justice to logic as he saw it. He tried to convince the administration to appoint at least one other logician of his stripe, but he was not a successful politician in the academic world. The dispute took its toll in personal animosities, resignations, and disappointments. After a time he gave up trying.

Van Heijenoort's daughter Laure was an undergraduate at Brandeis during her father's tenure there. They had lunch together at least once a week and enjoyed the opportunity to interact on an adult level. He considered her presence as one of the great fringe benefits of his new job. Laure later reported that at a certain point her father seemed to have lost interest in teaching. The enthusiasm was gone. He stayed on at Brandeis until his retirement in 1977 and, without fail, fullfilled his obligations there, but his passions were elsewhere.

* * *

*From Frege to Gödel: A Source Book in Mathematical Logic, 1879–1931*, edited by Jean van Heijenoort, was published by Harvard University Press in 1967 as part of its History of Science Series. Although by the time van Heijenoort was appointed as a professor at Brandeis he was well-known and respected in certain circles, the publication of the *Source Book* and the acclaim that followed broadened and enhanced his reputation. Without quite realizing it, van Heijenoort had been collecting material for such a book long before he was asked to be Harvard's editor and before he came to Brandeis and the Cambridge area. He was a persistent autodidact, and from his early dabbling in *Principia Mathematica* at the Forty-second Street library he became interested in the foundations of mathematics. Then, spurred on by the inspirational encounters he had had at the logic meeting in Cornell in 1957, he began, systematically, to wade through the literature in the field of modern logic. Not everyone who works in a subject thinks it necessary to know all the work that has been done in that subject, but van Heijenoort's approach was total immersion.

It was not an easy task he had set for himself; many of the papers were not readily accessible for reasons of language or location. However, van Heijenoort persevered, and as he did so it struck him that the work he was doing would be of value to others. While he was still at NYU, he began to think of putting some of the most important papers into one volume. Coincidentally, Professor Willard Van Orman Quine of Harvard was asked by Harvard University Press to suggest someone to edit a source book in logic for a History of Science series. Quine did not know van Heijenoort, but he had heard of him through his colleague, Professor Burton Dreben, who knew exactly what van Heijenoort had been doing and knew that he was ideally suited for the job. In 1959, van Heijenoort was invited to Harvard to meet Quine, after which Quine arranged a meeting with Harvard Press, and the book was launched.

Van Heijenoort was eager to take on the project but grossly underestimated the amount of work involved: six months, he had thought, and then found himself working "terribly hard, under heavy pressure, for seven years." The pressure was his own. He wanted to finish quickly, but felt he had to do an excellent job. In a frank avowal of the recognition he wanted, van Heijenoort later said, "I certainly didn't want to do anything sloppy; it was a question of prestige."

Beginning with the landmark monograph, *Begriffsschrift* (ideography or concept writing) by Gottlob Frege, and ending with the paper, "On the Consistency of Arithmetic" by Jacques Herbrand, van Heijenoort undertook the task of presenting forty-six papers, written in the period 1879 to 1931, in such a way as to give the reader a comprehensive picture of the evolution of modern logic. Among other things, the fact that the papers were written in seven different languages presented the editor with a serious problem. Its successful resolution was a major factor in the success of the book. In his preface to the first edition, van Heijenoort discusses the difficulties he faced in dealing with the early papers:

> Some authors were not writing in their mother tongues; some papers were hastily written, and many have a fair sprinkling of ambiguities, minor errors, and misprints; before 1930, the standards of editorship were not what they are today. All this means that the translators and editor had their share of problems. They strove above all to produce a clear and unambiguous text, but they have endeavored to follow that narrow path where accuracy combines with good style.

Although van Heijenoort was the sole editor of the *Source Book* and responsible for all decisions about the texts, he did have a distinguished group of contributors and translators. Professors Burton Dreben, W.V.O. Quine, and Hao Wang gave advice in the selection of texts, and they also wrote introductory notes

*Jean and Maman at Orly Airport. 1960.*

to some of the chapters. Professor Charles Parsons contributed a very important introduction to the Brouwer text. Besides these colleagues, van Heijenoort had help with translation from Stefan Bauer-Mengelberg, his former student from NYU, Professor Dagfinn Føllesdal, and Beverly Woodward, his student and girlfriend. The last paragraph of the preface to the *Source Book* reads: "As the reader will see, a large share of the translations from the German devolved on Mr. Bauer-Mengelberg. Moreover, he went over the whole manuscript and suggested many changes for the better." Over a period of six years, Bauer-Mengelberg did indeed devote an enormous amount of time and energy to van Heijenoort's project. Stefan's story of how and why he accepted the assignment reveals as much about their close friendship and personal characteristics as it does about their work habits and scholarship:

> Just after Van came back from India [where he had been a Fulbright Scholar in 1960], I met him in East Hampton at a dinner party at Margie Jonas' house [Margie Jonas was a mutual friend]. He was being his most charming and expansive self that evening.

*Van, the lobster cook, at Marjorie Jonas' house. East Hampton, 1963.*

Sometimes at a party, he would sit and not say a word, and one didn't know whether he was smoldering about something that had made him angry or what, but in any case he would be tuned out. But that evening he was in a rare mood, vital and vibrant, talking about Ceylon, raving about the beauties of the Indian landscape, telling about what it was like to go to one of those long Indian movies.

Then, he brought up the subject of the *Source Book* he had begun to work on and said he would like me to do some translations. I said, "Really, I'm not interested," because at that time I was off on my musical career and had just finished a year with the New York Philharmonic and was conducting in Saint Louis. I was thinking of *that* sort of life.

Bauer-Mengelberg's uncle was Willem Mengelberg, the well-known conducter of the Amsterdam Concertgebouw Orchestra. For years Stefan had vacillated between an academic career and a concert career, often doing both at the same time. He had studied and worked with Leonard Bernstein, Carl Bamberger, William Steinberg, and others in the United States, and abroad with Nadia Boulanger in Paris and Willem Otterloo in the Netherlands. He had been guest conductor with at least half a dozen major orchestras, and also had been a voice coach at Sarah Lawrence, the Juilliard School and the Mannes College of Music. Later he would become president of the Mannes College for three years.

Van Heijenoort knew all the details of Stefan's musical career, but that did not stay his pursuit. As Stefan explained:

> When I told Van I wasn't interested, he said, "You know, you will be very well paid."
>
> "What did you have in mind?" I asked.
>
> He said, "One cent a word."
>
> I asked if he had a second argument, and he said, "Yes. Think of the *gloreee*."
>
> "Van," I said, "I am presently looking for *gloreee* in other fields."
>
> At that point he brought out his last argument, "Well, there are some translations which I just cannot give to anyone else," to which I said, "Flattery will get you nowhere."
>
> Of course, it got him exactly what he wanted. A week later, "by coincidence," I met him again at Margie Jonas's apartment in New York, and "just by chance" he had the table of contents with him and he showed it to me. When I saw the list of famous names and great papers, I could not resist. I really wanted to come to grips with that material—to make it accessible to others, and I knew that it would be very agreeable to work with Van. To make a long story short, I agreed to do ten translations for him, but in the end I translated twenty-eight out of the forty-seven papers in the book and worked on them pretty much full time for six years.

"Full time" for Bauer-Mengelberg did not mean that he quit his other jobs and devoted himself exclusively to the texts. Stefan always managed to do at least two full-time jobs. In that period, simultaneous with his musical career, he was teaching at IBM Systems Research Institute. Since he was using some of the *Source Book* materials in his courses, IBM was willing to foster the project. In the summer, they allowed him to leave Thursdays at noon, and to return Tuesdays at noon and to arrange his teaching schedule on the days he was in New York. Thus, four days a week, in the months between April and

*Stefan Bauer-Mengelberg, with Beethoven score. ca 1960.*

October, he would stay with Van in his country place in The Springs, and work on translations.

On the surface, it was a marriage of opposites: Stefan, stocky, ebullient, talkative; and van Heijenoort, tall and silent, sometimes to the point of withdrawal. But their work habits and their attitude toward the task at hand were the same, and they got along extremely well. Although Stefan had come from Germany to the United States as a boy and had quickly become Americanized, his early schooling in precision stuck:

> We were both equally compulsive about that last comma so we didn't argue those points. In fact, we never had any problems, not even about whether or not we should go to the beach. We had a very definite routine. On Thursday evening we would plan our four days of work; early Friday morning we would go shopping for food, and then we would settle down to the translations. There were times when we would spend an entire afternoon on one sentence. Of course, not all of it was done that way or we would never have finished, but when we came to something difficult, we decided we could not think of clocks as ticking on. We had to adopt an almost oriental attitude towards time, and I think it paid off in the end. It was sort of a relentless search for perfection.

Sidney Morgenbesser, who sometimes came to visit in the *"Source Book"* years, later described the atmosphere of the house in East Hampton as "something between a country retreat and a research center." At one time or another the four or five large tables in the living room had been covered with all the parts, pieces, and variations of the forty-six papers in seven languages which comprised the *Source Book in Mathematical Logic*. It was, most likely, the ghost of these papers that had prompted Joseph Stillman, the man who later bought Van's house, to see it as, "a place someone had come to *do* something–not just to be there."

In his understated fashion, Van was very proud of having produced the *Source Book* and pleased by the almost universal praise that was bestowed upon him when the volume appeared in 1967. If, in his own mind, he later diminished the value of the work somewhat, it was because it was "only history." Like all logicians and mathematicians, he harbored dreams of breaking new ground. But meanwhile, in the *Source Book,* he had marshalled his considerable talent for organization and analysis to break new ground in the way of standards and presentation of historical material. He had done something very valuable in the service of others, and he did take satisfaction in the achievement.

Pros and beginners alike were grateful for the *Source Book.* W.V.O. Quine, who had recommended van Heijenoort to Harvard University Press, had this to say:

> He made a contribution to mathematical logic that was second only to what Alonzo Church had contributed in molding the *Journal of Symbolic Logic* itself. ... For us it was a boon just to have these papers brought together, to have xeroxes in our hands of the original German. But then, he did these painstaking translations and painstaking commentaries ... a collaborative job, but very largely Van's own work.

The philosopher Judson Webb first read the *Source Book* when he was a philosophy graduate student just beginning his studies in logic at Case-Western Reserve in Cleveland, Ohio. His reaction then was simply that he felt lucky to have come across exactly what he needed, when he needed it. It was as if the book had been written for him. Webb later recalled his first encounter with the material in the text:

> Our library did not have all those original papers which he had dug up out of obscure European journals. I would have been completely incapable of doing that for myself at that time. The book was a lifetime of labor, I thought. All you had to do was look at it casually to see the overwhelming amount of work it must have taken. In fact, it made me tired just to look at it—to feel the effort that went into it, the care, the years of scholarship and reflection on these matters, because not only did he translate all the articles, but he also dug up the even more obscure sources on which all the papers were based, and then he wrote introductions to these very difficult articles. It was a gift, an act of generosity, not only for students like me but for people who were working on the cutting edge of the subject.

Judson Webb's early encounter with the original papers in the *Source Book* played no small part in his entering more deeply into the field of logic. As he made more serious use of the material, his understanding was enlarged and his admiration grew. He realized that van Heijenoort's work offered rewards far greater than mere convenience; in fact, it provided a full picture of the creation of modern logic. A few years later, Webb was hired as an assistant professor at Boston University, and, in due course, he met van Heijenoort who was then teaching at Brandeis University. He was delighted to find the older man accessible, generous with his time, and willing to answer his many questions about the history of logic. They met frequently, talking always about their professional and academic interests. It was close to two years before Webb discovered there was more to know about his colleague:

> I heard a couple of rumors from my friends. "Oh yeah, van Heijenoort," they would say. "Did you know, he used to be Trotsky's bodyguard?" And they would add a few other bits and pieces of his story. When I learned that he had lived with Trotsky for seven years and that he had worked full time in radical politics for thirteen years, I just couldn't believe it. I was astounded; absolutely dumbfounded. Not that I doubted it, but I couldn't fit it in with the rest of him; so mild, so cautious, and I couldn't see how he had had the time since I still had the idea that the *Source Book* had to have been done by someone who was thinking about these things from the cradle.

Anyone who ever worked with Jean van Heijenoort was aware of how much attention he paid "to the narrow path where accuracy combines with good style," as he put it in the preface to the *Source Book*. Solomon Feferman, with whom he would work on the multi-volumed *Collected Works of Kurt Gödel*, found Van's "passion for exactitude" a blessing tinged with a curse. Van often brought surprising heat to their consideration of punctuation, style, and typography, turning these discussions into the kind of full-fledged arguments that usually occur over more substantive matters. The final result however, once the volumes were published, was that they were models of what such texts can be.

Chapter Fourteen

# The Archivist in Spite of Himself

1940–1986

I never wanted to be a historian because, as Henry Ford said, "History is bunk." It's new ideas, invention that changes the world. But, because I was mixed up with history at a crucial time in the Thirties, I feel I understand things that many people don't understand today, and, then, there is the pedantic side of me that wants to establish the truth.

JEAN VAN HEIJENOORT, *1984*

WHEN LEON TROTSKY was expelled from Russia in 1929 and sent into exile in Turkey, he had neither choice nor control over what happened to him and his family. Considering the dire circumstances of his departure and the perilous path of his exile—from Moscow to Siberia to Turkey to France to Norway and finally to Mexico—it seems nothing short of miraculous that, while he was on the run, he and his comrades succeeded in packing, crating, and transporting mountains of important papers and letters by train and by ship. Furthermore, each time he fled from one place to the next, there was more to move.

During his twelve years of exile, up to the very moment of his assassination, Trotsky wrote like a man possessed: many books, including his autobiography, *My Life,* a three-volume *History of the Russian Revolution,* and a biography of Stalin; hundreds upon hundreds of articles for newspapers, journals, reviews, and magazines; and thousands of letters. Copies and originals of all these, as well as articles and letters received, were duly filed into the mushrooming archive.

Almost from the moment young Comrade Jean van Heijenoort came upon the scene in Turkey, he, more than any other individual in Trotsky's inner circle, was the one who would be carefully placing documents into or out of the files. Trotsky, the teasing nomenclator, who bestowed a half-dozen names on his youngest disciple from "Comrade Engineer" to "Comrade Foreign Minister," eventually added "Comrade Archivist" to the list. Agreeing with Trotsky a half-century later, Professor Pierre Broué, an authority on Trotsky's exile period, said,

> No one understood the importance of Trotsky's papers and personal letters better than Van, not even the Old Man himself.

> When Van was absent for one reason or another, Trotsky would say, "Van isn't here so the papers are in disorder." But otherwise, Trotsky never paid sufficient attention to his archive until he had to use it to defend himself against the accusations of the Moscow trials.

Except when he was writing a historical book, Leon Trotsky seldom referred to his earlier papers or made use of his working files. Relying on his incisive wit and prodigious memory, he wrote most of his letters and articles off the cuff in an exuberant, free-flowing style. He loved the act of writing: the right phrase, the pungent comparison. There is no question that in spite of the hardships he endured, he positively enjoyed that aspect of his exile which gave him the opportunity to use his extraordinary literary talent with much more freedom than he would have had "at home."

However, when Stalin accused him of treason and conspiracy, his *papki*, as the papers were called, took on new and crucial importance as material with which to counter Stalin's charges. Although no compilation of even the most incontrovertible evidence was likely to alter Stalin's trumped-up charge of conspiracy, the existence of the archive and of documents by which Stalin's accusations could be refuted made it possible for Trotsky to garner sympathy and support, at least in certain circles. Without the presence of such concrete evidence to substantiate Trotsky's claims of innocence, it is doubtful that the Dewey Commission would have come to its conclusions in his favor. Throughout this critically important hearing, Comrade Van was a master at marshalling documents, finding just the right bit of evidence to do the job. Aided by Jan Frankel (who left Trotsky and the inner circle immediately after the Dewey Commission hearings were concluded), Van played an absolutely crucial role.

Nature and the French school system had endowed van Heijenoort with every talent and attribute of the historian, archivist, and scholar. He was patient, persistent, probing, interested in minutiae as well as the larger picture, meticulously careful, organized, orderly, hard-working, deep-digging, compulsive about finishing what he started, and unstinting in his help to others doing research in any area with which he was familiar.

*van Heijenoort in the Fefermans' living room. Stanford, California, 1984.*

But although van Heijenoort was fascinated by history and, since his school days, had read its pages extensively and thoroughly, he had no desire to be a historian. At no stage of his life was he interested in making that his career. History dealt with what had already come to pass; he preferred to think of himself as an innovator. "New ideas, *fresh* ideas, that's what I like," he said, and he said it often. His first passionate, quixotic choice had been nothing less than the idea of changing the world, or at least part of it, by revolution. That had not worked out too well. After thirteen years, with considerably lowered expectations, he had returned to his early love, mathematics, and later, philosophy, hoping to break new ground, or to find connections between these fields on the borders of mathematics and philosophy. Ironically, and to his regret, even in the small world of mathematical logic, it was not in the presentation of new, earthshaking ideas that he made his major contributions. Rather, it was precisely in the *history* of the subject that he ended up doing his definitively classical work.

While Van was sometimes embarrassed by what he called his "pedantic side," and said, apologetically, "maybe some people think I'm too compulsive," on balance he considered his drive for perfection a virtue. As a scholar, he made good use of all the qualities associated with that aspect of his character. Perhaps just because his personal life was governed by impulse and emotion, when his organized, professional side was engaged, he was all control and discipline. In his seventies, he acknowledged that doing a careful, painstaking historical task, whether in politics or mathematics, was, for him, a form of salvation. Keeping his nose to the grindstone involved his mind as well, and released him from the contemplation of his constant, seemingly unsolvable, personal dilemmas. Furthermore, this was the kind of work that yielded actual, visible progress, whereas his emotional life seemed to be an endless rerun of the same old story: boy meets girl, falls immediately and hopelessly in love, marries impetuously (or does something equivalent), repents, separates, and the spool rewinds and begins again.

* * *

In 1936, the purge trials and Stalin's renewed venomous attack upon Trotsky made it frighteningly clear to the Old Man and his followers that now, more than ever, their lives and property were in grave danger. In Paris, a part of the archives left for safekeeping with Trotsky's son, Liova, were burglarized. In 1938, Trotsky's former secretary, Rudolf Klement, the hapless comrade who had precipitated the trouble in Barbizon, was found floating in the Seine, beheaded. That same year, Sedov himself died under mysterious circumstances, and in Spain, Erwin Wolf, Trotsky's secretary and Van's comrade during the Norwegian misadventure, was assassinated.

Convinced that Stalin was bent on obliterating all traces of the Trotskyite opposition, written words as well as human bodies, Trotsky, in Coyoacán, proposed a sale of his archives to an "American institution," with the condition that the papers be guarded against the Stalinists and their agents. In March 1939, he wrote to Harvard University Library, the Hoover Institution at Stanford, and the University of Chicago Library. All three institutions promptly answered that they were indeed very interested, and all three sent representatives to Mexico to view the papers and present offers. Van Heijenoort, in the double role of Comrade Archivist and Comrade Salesman, was there to show the wares. He later recalled that at first Trotsky seemed to be leaning in the direction of Hoover; in the end, however, the quickest and most lucrative response came from Harvard and the deal was done.

Before van Heijenoort left for the United States, in November of 1939, he was involved in the packing of the *papki* for shipment. Included were all papers written before December 31, 1936, a date chosen because Trotsky had been at sea on that day, prior to his arrival in Mexico; thus there was a natural break. The rest of the archive remained in Coyoacán as part of the active and still growing files.

In the fall of 1940, only a few months after Trotsky's assassination, van Heijenoort, still stunned, still mourning, made his first visit to Harvard to open the crates and check the contents of the folders he himself had packed. There he found a second, unexpected shipment awaiting his inspection: the papers which Natalia Ivanovna and her lawyer, Albert Goldman, had sent off in haste, immediately after the murder, without a purchase agreement, in the hope that at Harvard they would be safe. In these boxes, along with Trotsky's Mexican residence permit and his old Turkish passport, van Heijenoort found the working files, containing the most recent material—items which he remembered filing up until the moment of his departure.

Now, these papers, which he had not expected to see again for years, were consigned to history, and there would be no more. *Peró's* inexhaustible pen was dry. For Comrade *Uzhe,* it was the final, incontrovertible confirmation of Lev Davidovich's death.

Following his initial 1940 visit to Harvard Library, which was sponsored by the Socialist Workers party of New York and encouraged by the library staff, van Heijenoort made several trips from New York to Cambridge to examine the contents of the archive and confirm that everything was indeed there. Then, in 1942, after the United States joined the Allied Forces in Europe, the Trotsky collection was suddenly declared off limits. As if it were a deadly virus which if released might do untold damage, the archive was sealed and labeled top secret. No one was permitted access to it without approval from the highest authority. Van Heijenoort, with lingering bitterness, called this action *une petite gentillesse* (a small kindness) granted by President Franklin D. Roosevelt to keep Stalin mollified while the United States delayed opening a second front on the European battlefield. Although it is not clear what his sources were, Van was convinced that it was President Roosevelt personally who saw to it that nothing critical of Stalin was allowed to surface while the war was going on. To support this contention, he cited the suppression of Trotsky's book on Stalin, which had already been printed and was ready for distribution when it was held back, and the trial of the leadership of the Socialist Workers party in Minneapolis, where eighteen Trotskyites were convicted and sent to federal prison for advocating the violent overthrow of the United States government. All this, Van said, was done to please Stalin and assure him that the United States was a true ally.

The Trotsky Archive remained closed until well after the end of World War II, and it was not until the early 1950s that van Heijenoort returned to Houghton Library at Harvard, this time as a consultant and expert in special acquisitions. By then, alliances had shifted, and the Cold War between the United States and the Eastern bloc countries removed former constraints and, in a way, served to encourage the opening of the archive. The political climate and indeed the whole world had changed, although not in the direction Trotsky or his followers had anticipated. If at one time there had been a tactical political reason to seal the archive, there was now every reason to organize it, to catalogue it, and to make its contents accessible to historians, biographers, and whoever else might have an interest in it. (There was still one remaining restriction: In the terms of the original sale, Trotsky and Natalia Ivanovna had stipulated that all the letters were to remain sealed for forty years, in order to protect the safety and privacy of the various correspondents.)

The world had changed and so had van Heijenoort. The man confronting the documents of his political past had struggled through an enormous transformation. The ex-secretary-bodyguard was now an assistant professor of mathematics at NYU; politics was no longer his *raison d'être*, Trotsky no longer his spiritual father, and Marxism no longer his ideology. He was possessed by another set of priorities and goals that related to an abstract, academic existence. The whole enterprise of his political life now seemed to him to have been a terrible mistake, and he sought to put it behind him, to lose and perhaps find himself in closely reasoned intellectual pursuits. And yet, when he shed his revolutionary skin, he did not—perhaps could not—shake loose from the history and the person associated with it. Whenever he was asked for advice or direct help by librarians, historians, students, or simply interested parties, Comrade Archivist-Professor was there, ready to point the way as guide and interpreter of the Old Man's papers.

Most often those whom van Heijenoort helped were still Trotskyites, or much more sympathetic to Marxism than he was, but that in itself did not diminish his willingness to be of assistance. There was no one who knew more about the archives or their creator than he, no one who had the proximity and, at the same time, the neutrality; for unlike many other former radical revolutionaries, van Heijenoort did not react with sharp bitterness and rage and roll over quickly from left to right. Trotsky had been *his* Old Man, there was never any question of that. Now he was a fallen idol, but disillusioned though Van was, he did not want to rob Trotsky of his heroic stature or his brilliance, or to deny his charismatic charm. To do so would cast a longer shadow than he wished on his own earlier commitment. On the other hand, as a scholar, van Heijenoort felt obliged to offer all that he knew about the history of particular events, as well as the details of everyday life with his leader, including the blemishes, self-indulgences, and mistakes. As for the archive itself, as in all things, he had an intense desire to be exact, to correct the mistakes others had made, and to insure that the record was as complete as possible.

The first person to ask for Van's advice and help was Douglas Bryant, who came to Harvard as assistant librarian in 1952 and was to become professor of bibliography and head of all Harvard libraries. Bryant, a man with broad international concerns, knowledgeable in history and foreign affairs, was particularly interested in the Trotsky papers. In 1932–33, as an undergraduate student, he had spent a year in Munich and had seen Hitler's rise to power. At that very moment, van Heijenoort in France, impelled by the same event, was deciding to give up mathematics and devote himself entirely to politics. While Comrade Van was following Trotsky in exile, Bryant completed his undergraduate studies at Stanford in 1935, and got his first job: part-time translator for the Hoover Institution at twenty-five cents an hour. (As coincidence would have it, fifty years later, van Heijenoort was also to do some important part-time work for Hoover, on *their* Trotsky Archive.)

For two years before coming to Harvard, Bryant had been in the foreign service, as an attaché to the American Embassy in London. When he joined the staff at Harvard Library, his first specific assignment was to get the last of the post-1936 Trotsky archival material out of Mexico and into the library. These were papers that Natalia Ivanovna had gathered from various sources after the end of World War II. It was for that purpose that Bryant called upon Trotsky's former secretary.

The two men, almost the same age, took to each other immediately. In spite of their vastly different backgrounds, they had certain gentlemanly qualities that were very much the same. Van Heijenoort admired Bryant for his intelligence and integrity as well as his professional competence. Bryant, like everyone else, was fascinated by the contrasts in van Heijenoort's makeup; his quiet, retiring manner, so difficult to square with his adventurous life, his courage and commitment. He was also impressed by the breadth and variety of his colleague's knowledge. "Any conversation with Jean was stimulating, exhilarating," Bryant said later. "One never left his company without feeling that one had learned from him. He had this limitless store of tales of a world far removed from my own." Reveling in the openness of the librarian's appreciation, van Heijenoort gave him the gift of his confidence. He paid Bryant a supreme compliment by speaking with rare candor about his personal experience with Trotsky, and with Marxist politics in general, at a time when he scarcely breathed a word to anyone about those subjects. The fact that Bryant had been in Germany in 1933 and not only understood but had directly observed the political circumstances in which Hitler had risen to power meant a great deal to van Heijenoort. The friendship had a life of its own and went far beyond the confines of their professional collaboration.

When Douglas Bryant spoke at the memorial gathering held in Jean van Heijenoort's honor at Harvard University in April 1986, he went into a lovingly detailed appreciation of his colleague's nearly forty years' work as consultant, bibliographer, and expert in special acquisitions:

> Jean's contribution to the Harvard University Library and thus to twentieth-century historical scholarship was, quite literally, incalculable. ... He alone organized and directed the immense job of cataloguing the vast and complex archive of Leon Trotsky which Harvard had acquired in two parts. ... After the death of Trotsky's widow, Natalia, the library acquired her own papers, which Jean then put in order, in association with her husband's files. To this monumental task he brought knowledge of people and events and places which he alone had. Thus, for instance, he was able to identify accurately the dozens of individuals represented in the archive, people who wrote and were referred to under countless pseudonyms. No one else, it is fair to say, could have disentangled this complex web and thus have left a priceless legacy to history. The innate tact and sensitivity, so much a part of his personality, were never more evident than on a visit Jean and I made to Mexico City to discuss with Madame Trotsky the very delicate questions concerning access to her husband's archive, and

the eventual placement of her own papers together with his, here at Harvard. These same characteristics of Jean's were many times brought to bear in negotiations we both conducted in this country and in Europe toward acquiring the papers of a number of Trotsky's associates. His patient persistence in these matters, over many years, had the happy result that he was able to bring to the library here the personal archives of at least half a dozen figures who were influential in the widespread international enterprise carried on by Leon Trotsky.

* * *

With considerable fanfare, the whirring of television cameras, and the presence of the international press, the exile correspondence of the Trotsky Archive was finally officially opened to the public on January 2, 1980. On January 7, Jean van Heijenoort, playing a rare role in the limelight, inaugurated a special exhibition of the letters in Houghton Library with a lecture which, according to *Harvard Magazine*, was delivered to "an audience of some three hundred rapt scholars, librarians, and friends of Trotsky." Also present was Estaban (Vsievolod) Volkov, Trotsky's grandson, who came from Mexico City, as well as many of van Heijenoort's personal friends.

Van Heijenoort opened his talk in the style of a travelogue, moving gracefully from beginning to end:

> We are here at the terminal point of a long trajectory. The initial point of that trajectory is Moscow, 1927. From there it passes through Alma-Ata, in Siberia near the Chinese border, then bends back to Prinkipo, a small island in the Marmara Sea; from Turkey, it goes on to Marseilles and Paris, then to Oslo and Hønefoss in Norway; from Norway the trajectory reaches Tampico, then Coyoacán, in the outskirts of Mexico City; and then finally, it goes from Coyoacán to Cambridge, Massachusetts.

From Prinkipo onward, van Heijenoort was describing his own journey as well as that of the papers; but never one to indulge in his own story, he concentrated on Trotsky and the significance of the letters, which had heretofore been unavailable. These letters, he said, would not produce startling political revelations:

> Rather, they add a new dimension. The published writings operate with political concepts, and it is not easy to feel the man behind the writings. Even Trotsky's autobiography, *My Life*, reveals little of his inner life. There is a lot of drama; we see Trotsky escaping from Siberia and racing across the steppes, but we don't know exactly what he feels. The correspondence is going to change that. Some letters are quite revealing about him as a man.

While van Heijenoort was focused on what would be special about the letters from a researcher's point of view, Rodney Dennis, curator of manuscripts for the Harvard University Library had some other very practical concerns regarding the opening of the Archive:

> After the collection was open to the public, all hell broke loose. *Everybody* came. It was the only time in the history of Houghton that there was a waiting line at the front door before the library opened. The reading room in January was as full as it ever gets in the busiest periods in July. There were historians, there were political people, people poring over great amounts of paper. Generally the work that researchers do here is on one particular manuscript, examining it with ultraviolet light, bombarding it with alpha rays, staring at it one way for six months and then turning it upside down. So, all of a sudden we have a lot of people who want to look at an *entire* collection. We had to move the papers from where they were to a place right near the elevator because they were being sent up and down continuously.

That was the scene after January 2, but before the opening the major worry among the library staff had been the security of the documents. As Rodney Dennis described it:

> We had this tremendous collection, catalogued right up to the nines. I don't think you'll find any other political papers anywhere else in the world in which there is an item by item account—but we didn't know *who* was going to be using these papers. What if people came in and called for manuscripts and just began eating them, right there in the reading room? For example, someone who had waited thirty years to make this stuff disappear.

Dennis was perfectly serious; he did not think the idea of someone "eating the papers" or using other drastic measures to wipe out the documents was at all far-fetched. Trotsky's assassination and the number of friends and comrades who had been "bumped off" were evidence enough of the genuine danger. The solution to this problem was to microfilm the entire collection, and this was accomplished before the doors opened on January 2. Then everyone involved with the archive relaxed, somewhat.

One last prickly thorn remained in Mr. Dennis's side: towards the end of his inaugural speech, van Heijenoort cautioned the potential users of the archive about mistakes in the catalogue and appealed to them

> to keep a record of any inaccuracy they find and give it to the librarian in charge at Houghton. We will devise machinery so that the catalogue can be improved by a kind of collective work. I must also say that, at the beginning, Mr. Miehe used, in good faith, Isaac Deutscher's book, and this book is notoriously deficient as far as dates, places, spelling of names ... are concerned.

Whereas the historian Pierre Broué laughed and labeled his colleague's punctiliousness "pure Van," Rodney Dennis was appalled that in an inaugural address Van would choose to focus on mistakes. The papers had been catalogued, under Van's direction, by Patrick Miehe, "a very gifted person," Dennis said later, and,

> He [Miehe] did a fantastically good job. Then, when it was all done, van Heijenoort makes this speech about all the mistakes ... and he didn't have the faintest notion that he had done anything wrong. He lives in a world where there are no mistakes; but what would *you* think and feel if you were the cataloguer, or if you were in charge of the catalogue, as I was?

But in fact, even in Van's world of logic where Dennis imagined "there are no mistakes," van Heijenoort found and pursued them like a fury. As Solomon Feferman (and others) have noted, "Van had the idea that it was possible to get everything exactly right. He never let anything go and sometimes it could just drive you nuts."

* * *

One of the unexpected results—a fringe benefit—of van Heijenoort's extended and extensive retrospective work on Trotsky's letters and papers was his own book, *With Trotsky in Exile: From Prinkipo to Coyoacán*, published first in France by Maurice Nadeau, in 1977, and by Harvard University Press, in 1978. He described it as "a book of recollections, which attempts to create the atmosphere in which Trotsky lived and worked during his years in exile." Completely in character, he also wrote it to correct the errors in "some writings that concern Leon Trotsky." In an appendix to the text, van Heijenoort devotes ten pages to the "correction of errors" in the work of Victor Serge and Natalia Sedova, Isaac Deutscher, Francis Wyndham and David King, Sara Weber, André Malraux, André Breton, and others. But the memoir served yet another, more personal, purpose: it gave van Heijenoort the opportunity to revive his own past, an enterprise that, not surprisingly, had a cathartic effect upon him. Here, finally, written under his true name, was a public avowal of his earlier, "other life" in politics.

Even so, of that seemingly intimate, first-person account, one might say the same thing about Van that *he* said of Trotsky: "It is not easy to feel the man behind the writings." Readers who come to the book with the anticipation of discovering, at last, the hidden side of the mysterious Jean van Heijenoort are often sorely disappointed. To be sure, in the last words of the book, when van Heijenoort writes, "Darkness set in," one feels the depth of his reaction to Trotsky's death and, in the afterword, he very clearly reveals his doubt and disorientation in the years that followed. So his book of recollections was an opening of a sort, an indication of a willingness to talk to the world at large about his years as a Trotskyite; and it invited questions, although not always those the author expected or wanted to answer.

Van Heijenoort frequently expressed amazement at how little most people (not just Americans) knew or understood of those times—meaning the 1930s in Europe—and conscientiously tried his best to explain. Thus, occasionally and tentatively, conversations were begun, but in true van Heijenoort spirit they were just as frequently squelched. It depended upon his mood and the style of the questioner. Professor Warren Goldfarb of Harvard tells of a German postdoctoral fellow who tried to engage van Heijenoort in conversation during a social evening at a meeting of the Association for Symbolic Logic in Marseilles with this opening:

> "Professor van Heijenoort, I understand you were Trotsky's secretary. What was that like?" Van Heijenoort looked slightly pained; he answered tersely with the dates of his time with Trotsky and the information that he had written a memoir about it; and then, before the postdoc could say anything more, Van immediately turned to me, addressed me formally as Mr. Goldfarb and asked me a very technical question. Clearly, the message was: I said what I wanted to say ... in the book; the last thing I want is casual (and uninformed) conversation about it.

* * *

Among those waiting in line at the door of Houghton Library on January 2, 1980, the official opening day of the Trotsky Archive, were Professor Pierre Broué, of the University of Grenoble, France, and Professor Marlene Kadar, then a graduate student in comparative literature at the University of Alberta, Edmonton, Canada. Broué, a historian, had already met van Heijenoort; Kadar, whose interest was cultural politics, had not. Both were eager to find material in the archive that would further their research; neither realized, at that point, the extent to which the original "Comrade Archivist" himself would contribute to and participate in their individual efforts.

In his earlier research, Professor Broué had come upon the name Comrade Van. He knew that Van had been one of the leaders of the European group, and he had heard various stories about him, including the rumor that he was dead. Broué attacked his own extensive work on Trotsky like a detective, tracking down all the present and former members of the multiple and various factions of the Fourth International and related movements. He had long lists of the comrades' names and aliases, their activities and involvements. After several unsuccessful attempts to locate Van, he concluded that the rumor of his death might well be true, especially since there was also a report that Van had gone to Russia and had been made a colonel in the Red Army. Ironically, that story was printed in *La Verité*, the journal young van Heijenoort once sold during his school days in Paris. "The people who published it claimed the story came from a Russian press release," Broué said later, and then speculated:

> Maybe it was an attempt by the Stalinists to encourage others to join them by using Van as an example of an important Trotskyite who did.
>
> I, personally, didn't know anything about Van except that he was supposed to have been one of the people closest to Trotsky, and that after he left Mexico and was in the United States, he testified at the trial of Mark Zborowski.
>
> Then, one day, Marguerite Bonnet, the person in charge of the Trotsky copyrights in Europe, called me and said, "Well! I have news of Van." I told her I thought he was dead, but she said, "No, he is a professor of logic in America and he will be coming to Paris very soon." I was very excited. I told her I had to see him, because I was working on a book about Trotsky and the Spanish Civil War.

The meeting between Broué and van Heijenoort, duly arranged by Bonnet, took place in the late 1960s at the Hotel Select, on the Place de la Sorbonne, in a room which Broué described as being "so small, two people could not sit down at the same time." The Select, right across the street from his former "prison," the Lycée Saint-Louis, was Van's regular pied-à-terre in Paris. Habit, nostalgia, and affection for his old haunts, if not his school, always brought the *taupin* back to the *Quartier Latin*.

The mood of their first meeting was strained. Broué found van Heijenoort cold, distant, and shy. Nonetheless Van gave him some valuable historical information, and he told him about the archive at Harvard. When Broué finished his book (a collection of Trotsky's writings on the Spanish Civil War), he sent it to van Heijenoort and waited for a response which never came.

"But much later, in July of 1984," Broué said, "he told me that it was after he received that book that he took me seriously."

It is completely out of character for the conscientious van Heijenoort to have failed to acknowledge receipt of the book; one can only imagine a letter lost in the post. In any case, once established, the contact between the two men continued through occasional meetings and correspondence, and by 1973, van Heijenoort was suggesting that Broué come to Harvard to make use of the sections of the archive that were already open. Broué did make one brief visit, but it was not until 1980, when the entire archive was catalogued and accessible to all scholars, that he began to work systematically and steadily on the material.

That year marked the beginning of a close professional working relationship, and a very special camaraderie blossomed between the two Frenchmen. Although their political perspectives were quite different—Broué, in large measure, was still sympathetic to Trotskyism—van Heijenoort could not help but appreciate and feel affection for a person who understood so much about the history of the man and the movement that had once been his life. The following two summers Broué came to Harvard, with a crew of assistants from Grenoble in his wake, to pore over and sift through the *papki* and to ask questions of the secretary who knew those papers in exquisite detail.

Broué's feelings about van Heijenoort bordered on veneration. Describing Van's work and personality, he said,

> Van amazed me; he should have been a historian because without any of the training, he knew all the tricks of the trade. In fact he was always telling *me* what to do. One day I told him that I had a list of a thousand Russian deportees and he said, "Oh, I thought of doing that, but I never got around to it." He had a very scientific spirit and an aptitude for all the techniques. He knew everybody's handwriting and style of expression and their personal attitudes too.

More than a friend, van Heijenoort acted as Broué's protector and mentor. He arranged for Broué's lodging in Professor Paul Doty's house on Kirkland Place in Cambridge (where Van himself lived), and he enabled Broué to make extensive use of not just the special collections, but the Harvard Library as a whole. In this regard, Broué revealed one more aspect of "pure Van" that had nothing to do with the latter's passion for correctness:

> I asked him if it would be difficult for me to get a card which would permit me to go anywhere in the library, especially the stacks. He said, "Difficult, no, but long, yes. I have a better idea." So, for three days we walked all over the library together, in every corner, speaking French in loud voices so that everyone who worked there would notice us. Then on the fourth day, I went by myself and everyone assumed I had a card. His plan worked.

Among the historians who made use of van Heijenoort's assistance, resources, and resourcefulness, Pierre Broué knew him best and benefited most. In an objective sense, he was excessive in his glorification, calling him "a universal spirit, an absolutely superior guy" ("*un type tout à fait supérieur*") and "the last Leonardo"; but there is no question that Broué was serious. He did not think he was exaggerating when he described Van as "capable of understanding all the problems that humanity is confronted with. He can understand almost everything, almost every intellectual discipline, and almost every psychological attitude, but at the same time, he has an interesting character himself, full of contradictions." Finally, when Broué exhausted his own lexicon of praise for van Heijenoort, he would say, "Did I tell you what André Breton said about him?"

Marlene Kadar, another early arrival at the archival font, also claimed Professor van Heijenoort as a mentor. She made at least four pilgrimages to Harvard while he was there, and he had a profound effect on her work, eventually becoming, in her words, "the revered external examiner for my Ph.D." Because of his complete and detailed knowledge of Trotsky's correspondence and his extraordinary memory of the contents of each letter in the archive, van Heijenoort contributed directly to Kadar's research on the triangle of influence which connected Trotsky's correspondents in Paris (André Breton and his followers), Mexico (Frida Kahlo and Diego Rivera) and New

York (the writers and editors of *Partisan Review*). In working with van Heijenoort, Kadar discovered that besides his inside-out knowledge of the archival material, he was "extremely well versed in modernist trends in American and European Art and sometimes he revealed his enthusiasm to me."

Dr. Kadar completed her dissertation, *Cultural Politics in the 1930s: Partisan Review, the Surrealists, and Leon Trotsky*, in 1983, and van Heijenoort was an external examiner at her oral examination at the University of Alberta. In this connection, she had *her* story to tell about Van's abhorrence of error and attention to detail:

> I remember the morning of my oral defense very vividly. He (Van) called me in to my supervisor's office, an hour before the exam was to take place, and, privately, he went over every page of the thesis, marking all the errors—even the most minuscule ones. He did it lovingly, perfecting it for "our" sake and for the sake of future researchers in the Trotsky archives. I admired him immensely for this, and I felt confident that the dissertation was finally correct.

This same profound attention to detail coupled with his astounding memory made it possible for van Heijenoort to provide Kadar with some totally unexpected material, only peripherally related to her thesis, which she was to use in her subsequent work on the correspondence of Earle Birney, Canada's poet laureate. Van Heijenoort remembered that a young Canadian poet, Earle Robertson, had visited Trotsky in Norway and corresponded with him afterwards, when the latter was in Mexico. About a dozen letters were exchanged. Thinking this might be of particular interest to Kadar because she was Canadian, Van brought these items to her attention. Indeed, it was of great interest. "Earle Robertson" turned out to be one of the many pseudonyms for Earle Birney, a poet Kadar had met and read several years earlier. In fear of the prevalent anticommunism of the times, Birney had used the pseudonym to conceal his identity. Close to fifty years had elapsed when Kadar presented the aged Birney with the facts of her discovery. By then it took Kadar a while to convince him that the documents in the Trotsky Archive were actually his own letters, for although Trotsky had saved all his correspondence to and from the poet, Birney himself had only a few letters. Eventually he recalled that he had buried the rest in a hill on the north side of Toronto and that when he went to dig them up after World War II, he had not been able to find them anywhere.

By a coincidence that would strain credibility in fiction, Dr. Kadar had the satisfaction of returning van Heijenoort's favor by bringing a stunning archival find to *his* attention. Toward the end of 1983, she came to Stanford University for a week, to peruse the resources of the Hoover Institution, to see what might be of interest to her. She had a second purpose: to call upon van Heijenoort, who was by then at Stanford, and to hand deliver a bound copy of her dissertation to him.

Early in the week, in the reading room of the Hoover Institution's archives, Dr. Elena Danielson, then an assistant archivist, showed Kadar a "home movie" from the Hoover film library starring Leon Trotsky, Natalia Sedova, Frida Kahlo, and Diego Rivera. There was also a tall blond man, a "bit player," in the picture, whom Kadar immediately recognized. "I know that man!" she exclaimed. "He's here at Stanford, right now—in the mathematics department."

It was, of course, van Heijenoort she had spotted. By 1983 he had become deeply engaged as one of the editors of Kurt Gödel's *Collected Works* and was spending at least half of each year at Stanford University working on the project. No one at Hoover, except perhaps Professor Sidney Hook (who after his retirement from NYU had become a fellow at Hoover), knew who Van was or was aware of his presence.

At Danielson's invitation, van Heijenoort came to the reading room and saw the film. Afterwards he, Danielson, and Kadar talked about which of Trotsky's papers were in the Hoover collection. Van Heijenoort said he was certain there was nothing of major interest, because in the recent past both he and Pierre Broué had looked through the catalogue and made several inquiries. Undeterred, Kadar encouraged Danielson to look for more Trotsky papers, and Danielson, on her own behalf, was eager to do so.

As she searched, Danielson noticed a new, *handwritten* entry for "Sedov" added to an old typed list of boxes and their contents. She was familiar enough with the various Hoover collections and their particular curators to recognize the handwriting and to deduce that the new entry must have been added at the request of Anna Bourguina, the curator of the well-known and extremely important Nicolaevsky collection. Bourguina had died the year before, but until then she had had complete control over who had access to those papers. Archivists and curators have a reputation for possessiveness toward their material that ordinary researchers consider strange and obstructive, but Bourguina's reputation in this regard was rare even among archivists. In fact, until her death, she had secreted the papers in her home, and the boxes had only recently made their public appearance. Therefore, Danielson selected those boxes designated by the handwritten note to show to Kadar partly because she herself was curious and interested to discover their contents. Of course she soon realized that the fact that they had never been catalogued would partially explain why van Heijenoort and Broué were unaware of their existence.

Kadar took a quick, random look at what Danielson offered her. When she spotted the signature "Van" at the bottom of a typed letter to Sieva, Trotsky's son, she stopped, dropped the document back in the box, and ran from Hoover Library to the mathematics department to tell that very same Van "to come and witness what I had found."

Kadar's two-block sprint was prompted by her memory of van Heijenoort's speech at the 1980 opening ceremony of the Trotsky Archive at Harvard in the course of which the professor, whom she had not yet met, had cautioned the assembled scholars about what he called "the holes in the archives" and in particular about the "big hole ... punched in 1933." He was referring to "the sensitive folders, those of the leaders of the French and German Trotskyite groups," that had been given to Liova Sedov for safe-keeping when the comrades arrived in France in 1933. Also among these papers, which van Heijenoort described as "very rich in content," was the extensive correspondence between Trotsky and his son, letters (written sometimes two or three times a week over a period of six or seven years) that revealed the rarely observed personal side of Trotsky. Dr. Kadar was almost positive that what she had just seen were precisely those long-lost Trotsky-Sedov papers which van Heijenoort thought had been stolen by Stalinist agents or taken by the French police at the time of Liova Sedov's mysterious death.

Exactly how and when the missing letters and papers ended up in the possession of Boris Nicolaevsky is murky. It is known that Nicolaevsky, an ubiquitous collector of everything relating to the history of Russian Socialism, called upon Sedov. Had he acquired them before the robbery? Had Sedov, in turn, given them to *him* for safekeeping? Or did Nicolaevsky take them from Sedov with a promise to return them and then "forget?" (In Pierre Broué's frank opinion, "Nico stole them!") Or did he gain possession of them after Sedov's death in 1938? One certain fact is that Nicolaevsky came to the United States in 1940, and that somehow or other, either before or after the war, or both, he got his enormous collection of papers to the United States by way of the diplomatic pouch and other means. He sold his "treasure" to Hoover in 1963, remained as the curator of the collection until his death in 1966, at which point the responsibility was inherited by Anna Bourguina, his long-time companion, secretary, and last-minute wife. (It is the conjecture of at least one archivist that he married her on his death bed in order to assure her succession as curator.) Why both Nicolaevsky and Bourguina sat on the Trotsky papers and kept their existence a secret, knowing full well they were thought to be lost, is, again, only a matter for conjecture and far beyond the province of this account.

Van Heijenoort was acquainted with Nicolaevsky; he had met him in Paris in the 1930s and had seen him in New York in the 1940s. He described him as an absolutely obsessed collector, related in acquisitive spirit to the protagonist of the Henry James story *The Aspern Papers*, in which the narrator comes within a hair's breadth of marrying in order to get hold of certain letters. "Read that," Van said later, "and you will understand Nicolaevsky. That man would go through your wastebasket. He would look at what was on your desk. If you didn't hold on to what you had—with both hands—he would take it away from you, saying, 'You don't need this. I'll keep it for you.' I never asked him if he had the Sedov papers. I didn't think he did, but if I *had* asked him, it would not have made one bit of difference because he would *never* have told me he had them."

Dr. Kadar brought the incredulous van Heijenoort to the reading room of Hoover Archives. In her account of the moment Kadar said:

> He was in a state of shock. I will never forget the look on his face when he saw the boxes of letters, boxes he had feared were lost forever. He became teary-eyed, and within a few minutes he closed up the files and left the library. He told me that he had to phone Pierre Broué right away. And he thanked me. I think he was amused that "a young girl from Canada," which was what he called me, had found the missing section.

The coincidence of the discovery meant a great deal to Kadar, although the papers themselves were of no particular importance to her work. She felt privileged to have played the part of catalyst in solving the mystery of the gaping "hole" that had troubled Van for so long. Had she not been at Hoover to encourage Danielson to "look for more," had she not heard van Heijenoort's earlier speech at Harvard, and had Danielson herself not been intrigued as well as perspicacious, who can say how long it would have taken for van Heijenoort to have knowledge of the contents of the Nicolaevsky boxes, and, therefore, for the Hoover archivists to understand completely the significance of the documents? Certainly the papers would have come to light in their own good time, but given the subsequent tragic events in van Heijenoort's life, it remains an ever-open question as to whether Trotsky's former apostle would have been there, ready to help decipher and interpret them.

But in 1983 he *was* there and so, once again, Comrade-Scholar Van was drawn into the role of decoder of the past, and this time one could hardly call him reluctant. Amazed to be actually holding the old familiar notes, letters, and manuscripts in his hands, he went through them, systematically, as was his wont. Thrilled though he was, after his first trembling scan of the Trotsky-Sedov papers, he refused Dr. Danielson's proposal that he take a quick, sampling overview of the entire collection that Nicolaevsky had gathered. Instead he told her that he would come back in a few days, and then he would want to see everything in the order in which it had been arranged by Madame Bourguina. He was given a desk and for several hours every day, in short takes, he revived and relived the years between 1931 and 1938.

Early in 1984, as soon as Professor Broué was able to leave his duties in Grenoble, he flew to Stanford to see for himself exactly what was in the new Nicolaevsky material. After van Heijenoort, there was no one more elated by the discovery of the missing link than he. It was more than a matter of the gap's being closed; as a biographer and chronicler of Trotsky's life in exile, Broué was especially qualified to appreciate the richness of the discovery. Even a cursory glance led him to believe that the letters between Trotsky and Liova were going to give further insight into Trotsky as a man, standing down from the political pedestal.

Broué's first trip to Hoover had been for a quick perusal only; he came again for a longer stay in the summer of 1984. By then, van Heijenoort had done a preliminary separation and classification of the papers, putting each one in a separate folder. However, the real work began when the two men, sitting face to face, set about identifying every single item in the collection—Broué with his wide-screen historical perspective and Van, the former secretary-intimate with his authoritative "I was there" knowledge of the personal and political context, including style, pseudonyms, code names, and so forth. He was the only person alive who still knew such details from first-hand experience.

Attached one minute, detached the next, Van sat with his colleague Broué and commented upon the illegibility of his own handwriting in those days, the idiosyncracies of his old comrade's prose, the tensions of the specific moment in question, and the brilliance and irascibility of the Old Man. Broué came away newly impressed by his colleague's incredible memory for detail, large and small, for his perceptiveness, and above all for his devotion to the task at hand, even when he had other pressing matters in the Gödel project to attend to.

As he had done in Cambridge, Massachusetts, van Heijenoort arranged for Broué to rent a room in the same house that he lived in. Thus, they spent a great deal of time both on and off the job discussing Trotsky, and Broué noticed what he thought was a change in van Heijenoort's feelings about the Old Man. As an example he described Van's reaction to a letter from Liova to Natalia Sedova in which Liova vented his feelings about his father. In part, the letter read,

> I think that all Dad's deficiencies have not diminished as he grew older, but under the influence of his isolation ... have gotten worse. His lack of tolerance, hot temper, inconsistency, even rudeness, his desire to humiliate, offend and even destroy have increased.

As Broué and Van read the letter together, Van, in complete agreement with Liova's assessment of the Old Man's character, nodded his head and muttered, "Naturally, naturally; of course, of course," but then he asked Broué, "Did Liova send the letter?" When Broué said no, Van responded by saying, "He did the right thing."

The point, according to Broué, was that, "Eight years ago, Van would have said: You see what I mean? You see what a bastard Trotsky was, what an *emmerdeur*? But now he says, it's good that Liova didn't send the letter."

Broué interpreted Van's changed response as evidence of a reconciliation with Trotsky. Whether or not that was true, it was clear that van Heijenoort had arrived at a more comfortable accommodation with his past. He enjoyed the irony of his presence at the Hoover Institution where he, *le vieux bolshevik,* mingled with the conservatives as a co-equal and was accorded great respect. He laughed all the way to the bank, as it were—the currency being legitimacy. The last barriers were down, and in those final years, although he still did not volunteer information without first being asked, he talked about his radical Marxist days with greater and greater ease, not just to his colleagues who already knew about it, but to almost anyone who posed the right questions.

# IV

Chapter Fifteen

# The Door to Life

Mexico: 1960–1980

*Benjamin and Jean Gollay, Jean van Heijenoort and Anne-Marie Zamora at Tenampa, "Mexico City's night life headquarters for the bohemians." 1975.*

It was a really big love story, at least for me. When I met him, I said to myself, this is one of those men that are born only every fifty years. He seemed to me like a flame. Pure.

Anne-Marie Zamora Laurent, *1984*

IN 1958, IMMEDIATELY after Jean van Heijenoort became a United States citizen, Douglas Bryant asked if he would accompany him to Mexico to help negotiate the purchase of Natalia Sedova's papers for the Trotsky Archive at Harvard. Van had already made plans for his long-awaited return to France, but with Trotsky calling from the grave, and Bryant pleading for his assistance, he agreed to delay his trip home. Thus the ex-Comrade Van went to Mexico in the capacity of "expert in special acquisitions" for Harvard University Library. This was, in Van's words, "an extremely delicate situation," requiring all the tact and sensitivity he could muster. For business reasons he would have to meet with Natalia, and for personal reasons he *wanted* to see Natalia; but what would that encounter be like, with the passionately loyal widow knowing that he had left the Trotskyist–Marxist fold and was no longer the ever-willing Comrade *Uzhe*?

Van Heijenoort prepared for the meeting by calling on Adolfo Zamora, and bringing Douglas Bryant with him. Zamora had been both Diego Rivera's and Leon Trotsky's friend and lawyer, and after Trotsky's murder, he had taken on the role of protector and advisor to Natalia Sedova. Van knew Zamora's support would be crucial to the acquisition of not only Natalia's papers but also of other documents.

As political comrades, Zamora and van Heijenoort had always liked and admired one another. There had been immediate rapport between them in the years when Van was Trotsky's secretary and Adolfo was a regular visitor at the blue house in Coyoacán. Zamora was intelligent, cultured, and "almost" French. He had lived in Paris, had studied law at the *Faculté de Droit,* and had married a Parisian woman. Because of his French intellectual training he, perhaps better than anyone else, knew what Van had given up to follow Trotsky, and because of his wife, he was attuned to French manners. Adolfo was awed by the young comrade's sacrifice and dedication, and Van was gratified to have a friend who had some understanding of his background, and spoke his language.

The visit with Zamora was a success; the lawyer approved of Harvard's offer and thought the transaction would benefit all concerned: the papers would be in a protected environment where they would be of use to scholars, and Natalia would receive some much needed cash. Assured of Adolfo's cooperation, van Heijenoort went to see Natalia Ivanovna and Estaban (Sieva) Volkov, Trotsky's grandson. Natalia and Sieva (and his wife and children) were still living in the house on Avenida Viena where Trotsky had been murdered, and where the bullet holes from the first assassination attempt could still be seen in the walls of the bedroom and study.

What a return that was for the ex-comrade turned professor. Nineteen years had elapsed since he had said goodby to Trotsky and Natalia, and there had been sea changes in his life. To Sieva, he seemed like a different person. "There are two Vans," Sieva said later, "the young Van, very charming, very active, very dynamic, very sympathetic, always solving problems; and the one I saw in Mexico when he came to get the papers: silent, sad, distant."

Although the reunion was subdued, the old comrades greeted each other warmly and tenderly, expressing almost familial feelings. But at another level the atmosphere was charged with unspoken currents of tension. Van knew that they knew he was a political renegade, but this fact was never discussed. Also, there was the residue of guilt he bore for having been absent when the Old Man was assassinated. Again, nothing was said; no one blamed him in any way; everyone knew Trotsky himself had rejected Van's offer to return to Coyoacán and to take charge of security. Nor could it be proven that his presence would have made a difference, but *he* had thought otherwise, and the guilt that was renewed in Trotsky's house was no less painful for being self-engendered.

In spite of all the troubling past history, the numerous meetings concerning the acquisition of Natalia's papers, and the question of access to Trotsky's papers and letters already in the archive went very well. In the end, Harvard acquired not only the widow's documents but also the so-called "satellite papers" which were the personal archives of a number of Trotsky's associates. None of this was accomplished in a hurry; it took many years and many trips to Mexico. But Bryant and van Heijenoort were patient men, and, as far as Van was concerned, these trips, paid for by the library, were a plus rather than a minus, a positive aspect of his work on behalf of Harvard. He liked to travel, and at that point he was delighted to be able to return to Mexico often.

Van saw Adolfo Zamora on every visit. As Natalia's lawyer, Zamora was advising her on all aspects of the negotiations for her papers; in addition Van and Adolfo discovered they had interests in common besides the professional matter of the archive, and they took pleasure in one another's company. Then, because Adolfo thought his daughter Ana Maria would find Van interesting, one day, as the archival business was drawing to an end, he invited her to join them. Years later, Ana Maria remembered the specific moment:

> My father called me at work and said that Van was in Mexico for the Natalia papers and that they were going to Cuernavaca. He asked me if I wanted to go with them and I said yes. And when I saw him, I was ... like that!

The Cuernavaca encounter was not Ana Maria's first meeting with Jean van Heijenoort. She had seen him often when she was a child and he was Trotsky's tall, blonde aide-de-camp. Her father had been in the habit of bringing her along on his frequent visits with Trotsky in Coyoacán, and Comrade Van had also gone to the Zamora house in Mexico City on business. Early on she had absorbed her father's view of him as a hero, a "pure man," who willingly devoted his life exclusively to service for "the cause." Long after Comrade Van left Mexico, Adolfo Zamora had continued to talk about him in those glowing terms.

A few days after Cuernavaca, van Heijenoort called Ana Maria at her office and asked if she would like to visit the pyramids with him, and she accepted. On his next visit to

Mexico, he called and invited her to lunch, but this time Ana Maria hesitated. As she said later, "I knew that if I went, something was going to happen; but then I said to myself: 'I cannot close the door to life,' and I accepted the lunch."

Anne-Marie (as Van always called her) was thirty-seven, a lawyer, with a ten-year-old son and an unhappy marriage. She was fair, with voluminous reddish-blonde hair, slender, and rather tall—not at all Van's "ideal type." Benjamin and Jean Gollay, van Heijenoort's old friends from New York days, described her as "very charming, very bright, very dynamic, and very different from all his other women." Ben Gollay added a reservation: "She was a dominating character, a little too harsh for me." Van's friend, Stefan Bauer-Mengelberg, concluded that she was not an exception to his own observation that Van had "a positive tropism towards kookie women."

What others thought or said out of earshot mattered not. Van and Anne-Marie were ready to fall in love, and they did. She had remembered everything her father used to say about van Heijenoort. Perhaps he had changed, but the person she was reintroduced to on the way to Cuernavaca did not disappoint her. As for Van, three marriages and several long liaisons had not dimmed his ardor or enthusiasm for romance. If he had lost his passion for politics, women still had their intoxicating effect upon him. When it came to affairs of the heart, van Heijenoort was unable to see more than two or three weeks ahead, or to imagine what a long-term relationship might be like. Nevertheless, this time, it was clear to him there were many positive virtues, among which was the fact that Anne-Marie was half-French and spoke the language like a native. For the first time since his separation from Gabrielle Brausch he had a woman to whom he could speak from his French heart. He felt he could tell her anything and she would understand. In this instance, he was talking about more than language. He thought he had found a soul-mate akin to his old friend Frida Kahlo, and the effect upon him was striking to those intimate few who saw them together.

Anne-Marie was one of the rare people who did not make a point about van Heijenoort's silences or reserve. "He would talk to me about things he never spoke of with anyone else," she said later. "He responded in a free way. I never found him silent." Laure van Heijenoort corroborated Anne-Marie's characterization of their relationship, saying, with wonder and amusement:

> When he was with her, he laughed out loud; he wore bright colored Mexican shirts and loud Hawaiian shirts—with flowers. He even told jokes in elevators. Can you imagine *my* father doing any of those things?

Although Adolfo Zamora was accustomed to his strong-willed daughter's impulsive, assertive ways, he was nevertheless taken aback when Ana Maria announced that she was going to marry *his* friend, Jean van Heijenoort. Still, he did not object to the marriage until his wife voiced opposition and then convinced him that it was a terrible mistake. Given Zamora's admiration for Van, it is not completely clear why he and his wife disapproved with such intensity. Their only stated objection was the twenty-year difference in age, a matter which did not concern the happy couple at all. Van Heijenoort was fifty-seven, but at that point in his life, according to his later account, he never thought about his age. As for Anne-Marie, she had found the love of her life, and nothing would persuade her to give him up. So Anne-Marie divorced her first husband (a simple procedure in Mexico if one is not a practicing Catholic) and married Van in Mexico City in 1969.

What other objections might the elder Zamoras have had? Perhaps they were worried, and not without cause, about their daughter's emotional instability and the effect this marriage might have upon her, for she had had some periods of serious depression during her first, unhappy marriage. Perhaps they feared she and her son might leave Mexico, or perhaps their objections had to do with money, because Anne-Marie was considered to be a wealthy woman. Whatever may have been the case, they were so angry that they refused to see or speak to their daughter for three years after she and van Heijenoort married. This state of affairs, and most particularly, her father's silence, troubled Anne-Marie deeply because she had always been very close to him.

Like her father, Ana Maria had studied law and became a lawyer, working in civil administration and business. It was a deliberate choice. Adolfo was her intellectual mentor; it was he who set the standards and values of education and culture in their family. Very early and very consciously, Ana Maria chose to identify not with her mother but her father. "I never had good relations with my mother," she said later, "although I came to understand her dilemma."

Señora Zamora, born Anne-Marie Laurent Bidault, had been a student of medieval history at the University of Paris when she met Adolfo, and she went with him to Mexico when his studies were finished. He became a successful and influential lawyer whose clients, besides Rivera and Trotsky, included leading members of the liberal Cárdenas government. Meanwhile, the transplanted Anne-Marie had three children, and the responsibility of running a large house. She did not have the opportunity or the drive to pursue her interests in medieval history. Ana Maria, the oldest child and her father's favorite, was aware that her mother was unhappy and resentful for having been cut off from her work. She saw her mother's frustration, but decided it was self-imposed. "She thought it was her duty to stay at home," Anne-Marie said later, "but it was not my father who made her think so." The daughter chose a different path. She would be an independent woman, liberated and assertive. Sieva Volkov, five years older, who after his grandfather's assassination was adopted by the Zamoras (Adolfo was his legal guardian) and lived with them for many years, remembered her as a "sweet little girl," but others, less generous, have said she was "a spoiled child," that her father doted upon her and was unable to refuse her anything. As an adult, she was restless and impulsive, and gave the impression that she thought her impetuosity was part of her charm. With pride she told an acquaintance of how she had taken an airplane from Mexico City to Veracruz just to buy an ice-cream cone there. By her peers she was perceived as a woman ahead of her time, a transgressor, liberated sexually and professionally; but that liberation had its costs. As one astute young woman observed, her gaze was often inward and she looked lost.

When Anne-Marie and van Heijenoort met and "opened the door to life," she was working as an administrator in the Mexican tax collection agency, the head of a department with forty employees. "I was the only woman in that kind of job then," she used to say, with pride. Besides that job, she ran a small, private kindergarten and dabbled in real estate sales. Not long after the marriage, she took leave, and she and Van spent the spring of 1971 at Stanford University where van Heijenoort had an appointment as visiting professor of philosophy. She made visits to the Cambridge, Massachusetts area while he was teaching at Brandeis, and to Long Island, to stay in van Heijenoort's house in East Hampton; and together they traveled abroad, to England and the European continent. Of course, van Heijenoort proudly took his new bride to his native turf to meet his family and to show her the battleground of his childhood. The family in Creil played their part and welcomed Anne-Marie as they had done all the other wives, lovers, and friends that their extraordinary wandering star brought home to them. At the same time they continued to maintain close relations with Van's second wife, Bunny.

Both Van and Anne-Marie later described the early years of their marriage in glowing terms, "like a honeymoon that went on and on." In a rare admission, which contradicted his usual claim that all he ever wanted was "one place to live, one job, and one woman," van Heijenoort said, "It added a lot of spice to separate and then to meet again. It was always fresh, always something new." Anne-Marie, a woman pleased to flout convention, was also happy with the rhythm and excitement of their life. She would say, "It was not a common marriage. It was not a common life." The travel to distant places appealed to her, and she liked seeing her husband in his professorial role at the many meetings and congresses to which they went. She admired him greatly, and in the earlier stages of their marriage, whatever he said or did was wonderful.

Between her visits to her husband in the United States, while he was teaching at Brandeis and working on the Trotsky Archive at Harvard, Anne-Marie would return to her home in Mexico City, to her son by her first marriage, and to her various business affairs. Before they married, the couple had never seriously discussed where they would live permanently, assuming that the matter would simply resolve itself, but when, in Anne-Marie's opinion, the time came to decide where they would settle and have the house she could call home, the resolution was far from simple.

If Van had undergone a transformation of sorts, not everyone noticed it. For most people, the barriers were still up. In 1972, during the same period that Laure van Heijenoort was marveling at his loud shirts and the jokes he told in elevators, Dr. Daniela Montelatici Prawitz, a philosophy student and later a psychoanalyst, met van Heijenoort for the first time at a professional meeting in France, which he attended without Anne-Marie. From that encounter Montelatici retained an image of "someone very pale, colorless almost, and so quiet. Still, there was something about him, something that made you want to know him." Indeed, this was the way most people in the academic world had always described him. And at the University in Mexico City, where van Heijenoort made new professional contacts and was invited to give a series of lectures on logic, it was the intimidatingly reserved, restrained, remote side of van Heijenoort that his students and colleagues encountered. Sieva Volkov's observation that "there are two Vans" seemed to be accurate. Outside the intimacy of his family, he had changed very little.

And yet, in 1976, at a summer logic meeting in Oxford, England, Jean van Heijenoort did something which others did regularly but which was quite extraordinary for him. He brought his wife to a social gathering of logicians. Now it wasn't that van Heijenoort didn't go to such parties. In fact, he actively liked parties, but usually he went alone. In general, his attitude was passive: a taciturn observer who moved from one group to another and occasionally made a lightly ironic comment. However, this particular summer evening, the atmosphere and his behavior was different. The host was Dr. Robin Gandy, a convivial fellow at Wolfson College. Gandy had a reputation for entertaining well, and the gathering was up to his usual standard, with plenty of good food and drink and lively conversation. Later in the week, Gandy would treat his colleagues to his annual Guy Fawkes Day fireworks display, but on this occasion the sparks came from van Heijenoort as he introduced Anne-Marie Zamora as his wife and told his friends that he was going to retire from Brandeis, and move to Mexico. He went on to say that he and his wife had a house in Mexico City and a house in Morelos, and anyone heading in that direction should be sure to visit them. All of this news amazed his listeners because, except for the inevitable circulating rumor of his Trotskyite past, the facts of his personal life had been a complete mystery. Scarcely anyone present knew before that moment that he was married, to say nothing of the fact that this was the fourth time.

Also, the open invitation to visit him and his wife at their homes in Mexico was something new. In the past, except in East Hampton, van Heijenoort had virtually never invited his colleagues to his personal residence. Stefan Bauer-Mengelberg attributed this to Van's "Frenchness," but others viewed it as part of his remoteness. Yet now, there he was, full of *bonhomie,* saying yes, yes, please come.

The house in Mexico City, near Chapultepec Park in the fashionable Lomas de Chapultapec district, belonged to Anne-Marie. On a small scale, it was a grand house, unlike any that van Heijenoort had ever called his. Both the residence and the neighborhood—where the streets are named for poets and philosophers and where most of the foreign embassies are to be found—were very elegant. Jean and Ben Gollay, van Heijenoort's old friends from New York, visited several times and were surprised to see the ascetic Van in such luxurious surroundings. Having been his neighbors in Long Island, they knew his house in The Springs, built and furnished in extremely simple if not spartan style. The house in Las Lomas, Jean Gollay later remembered,

> was all mulberry velvet and formal. The living room was a very serious parlor, with stuffed dark velvet furniture. The dining room had cabinets filled with tons and tons of stuff, crystal and dishes and tea sets and liqueur sets ... as if somebody had been a compulsive buyer. They had these heavy velvet draperies which were always drawn and the windows were all sealed; you could never open a window in that house.

To the Gollays, their friend seemed unexpectedly at home with all the plush and formality. Anne-Marie was an extremely hospitable hostess, constantly asking her guests what they needed or desired, prompting the servants to bring tea and cold drinks. "She kept those servants hopping," Ben Gollay said. "I wouldn't have thought Van could be comfortable or at ease in a place like that. It just wasn't him."

To all outward appearances, the old bolshevik had no greater problem than the next man accommodating himself to a life of luxury. Although he could easily and happily live in a monk's cell, he never insisted upon it. To his friend Stefan, he said, "You know, you want a glass of water. You push a button, and the girl, she brings it to you." But if luxury wasn't a problem, he soon discovered there were more general and larger troubles in his new environment to make him uncomfortable.

His list of difficulties began with the high altitude and was followed by the linkage of smog, pollution, automobile traffic, and noise that plagued Mexico City. By the early 1970s, well before the meeting in Oxford, he was already persuaded that he could not spend more than a week in any place of that altitude (7800 feet or 2380 meters) without drastic consequences to his physical well-being. He had headaches, he felt heart palpitations, and he could not breathe. In response to the obvious question of how he had managed to live in Coyoacán in the late 1930s for almost three years, he would simply answer that he had felt no ill effects at that time, but he did now.

In addition, van Heijenoort complained bitterly about what he saw as the disappearance of the unique qualities of Mexico City and the loss of the popular culture. He was especially upset by the disappearance of the neighborhood *pulquerías*, the little bars with brightly colored murals and decorations on the outside, where people would go to dance and drink. So, while on the one hand he enthusiastically issued invitations to his friends at the logic colloquium in England to visit him in Mexico, that very same week he had a quite different conversation with his student Felipe Bracho of Mexico City, who was doing his Ph.D work in Oxford. Bracho had attended van Heijenoort's series of logic lectures at the Universidad Nacional. These talks had stimulated the young man's budding interest in logic to such an extent that he eventually decided to study in that field. Later, when Bracho told Van he was attracted to the work of Professor Dana Scott, then at Oxford University, van Heijenoort encouraged him to go to England for his graduate studies and wrote a letter of recommendation that paved the way.

Van Heijenoort was demonstrably happy to see his former student in Oxford, and immediately proposed that they have dinner together. Felipé was momentarily taken aback by this unexpected effusiveness. Reporting on their evening together he said:

> I was very surprised, because, when I would see van Heijenoort in Mexico, he was helpful, but he was quite distant. Anyway, we had dinner together, and he had a little bit to drink—not too much—and he started telling me about how beautiful it used to be when he first came to Mexico City. He told me how he used to go to the *palenques*, the popular kind of theatre. Instead of a regular theatre, they would put up these tents, the *carpas*, for the performances. All the famous Mexican stars, especially the comics, started in the *carpas*. He said all that was lost, gone forever, and that now Mexico City was like any other big city, only noisier and more polluted. I was so moved, I was nearly crying, and I had never even seen the Mexico he was talking about.

Once Jean van Heijenoort came to the conclusion that certain conditions or actions were detrimental to his health, his conviction on the matter froze. The case was closed. Polite as ever, he would make a pretense of seeking and listening to other opinions, but he would quickly end any attempts at dissuasion with a slight smile, a deprecating nod, and a "Yes, yes, well, I know myself, and I know when something is wrong." Although various people, Anne-Marie among them, suggested that van Heijenoort's altitude sickness may have been encouraged by his general disaffection with Mexico City or by his general hypochondria, he would hear none of it, and in any event, it made no difference. As far as he was concerned, his symptoms were frighteningly real and something had to be done.

In Tlayacapán, a village not far from Cuernavaca, the van Heijenoort-Zamoras made their first and most serious attempt at a solution to the problem of where to live. There, on a

*The village of Tlayacapán.*

site with a spectacular view of plains and mountains, including the magical Popocatépetl and Ixtaccihuatl, they commissioned a father-and-son team to build a country retreat. In addition to its beauty, the location had other advantages: it was only an hour away from Mexico City, the altitude was two thousand feet lower, the air was clean, and there was only one paved road in the community, so noise could hardly be a worry.

The L-shaped house took three years to complete, with Ezequiel Valdés, the son, in charge of a crew of local men. It had a gigantic main room, with a large dining area at one end. There was a traditional Mexican country kitchen with blue and white tile counters and tile floors, and five bedrooms each with its own bath surrounding a jasmine-filled courtyard. Van and Anne-Marie were very much involved in all the details of design and construction, both practical and aesthetic. He particularly gave lengthy consideration to what would be the ideal cover for the large swimming pool in front of the house.

Laure van Heijenoort's description captures much more than the architectural grace of the house in Tlayacapán:

*The house Van and Anne-Marie built in Tlayacapán.*

*Sunset in Tlayacapán.*

> It was built of local stone and had some old wooden doors that looked as if they had once been part of a church. The room I stayed in several times had a uni-directional tile floor, but one tile was facing in the wrong direction, an indication that the person who laid the tile subscribed to the Indian belief that no one should ever do anything perfectly, lest the divinity determine that one's lifetime of striving should end. Outside they had a patch of corn, not far from the house, and there were some large exotic plants that Anne-Marie had brought back from Baja California. In the summer, the swimming pool was constantly full of live bees. Ezequiel's father had a net and would stand there for hours gently pulling the bees out of the water. It was a beautiful place.

When van Heijenoort married Anne-Marie in 1969 and made the commitment to live in Mexico part of the year he was, in a way, telling himself: you *can* go home again, and you can eat your cake and have it too. In the glow of his and Anne-Marie's "big love story," he had imagined he could recapture his Mexico of the thirties, the exotic land and the romance of the past, without the old burdens—without the obligations to Trotsky and the unswerving commitment to Marxism. Here was a way to remain connected to the virtues of the past in which he had been so deeply involved. With frequent trips to the United States, he could also continue his present scholarly historical-mathematical-logical life in the Cambridge–Boston area, and, if he desired, he could establish contact at the University of Mexico and be welcome there as a guest lecturer.

The move to Tlayacapán in the early 1970s was a search for the lost color, an attempt to live in the warmth of the old Mexico. And for a few years, van Heijenoort almost found what he was looking for. In that little village, where with his pale complexion, blond hair and light eyes he was the "gringoest" of *gringos*, he was as satisfied and as much at peace as his restless nature allowed him to be. While the villagers saw him as a silent, distant, wealthy outsider, he wanted and sometimes achieved a sense of participation.

As a step in the direction of belonging to the community, when their housebuilder Ezequiel and his wife Zenaida announced that they were going to have a child, Anne-Marie and Van asked if they might be the godparents. The Valdéses were confused by the offer since, according to normal protocol, *they* should have been the ones to ask; but they would never have done so because of the obvious economic and class differences between them. Ezequiel sought his priest's advice; the *padre's* answer was: first, think of the child and the obvious advantages to her, and second, if the *gringos* asked for such a thing, they must want it very much. Ezequiel and Zenaida said Yes, and Jean and Anne-Marie became their *compadre* and *madrina,* which meant there were bonds and loyalties between the two families that were much closer and more complicated than usual for house-owner and house-builder. Ezequiel and Zenaida continued to work for the van Heijenoorts after the house was built, and the old man Valdés and his wife lived in the caretaker's house and watched over the property when the owners were away. When the *gringos* were present, the whole Valdés family took responsibility for the welfare of their patrons and acted as their advocates in the village.

Padre Chavo, one of several priests in the village (but not Ezequiel's *padre*), had occasion to become well-acquainted with Ana Maria, as he called her. One day, one of his parishioners came to him in fear and trembling. He said he had been walking across *la gringa's* land, as he always did, when she came out with a shotgun, told him to leave immediately, and said she would shoot him if he ever set foot there again. The Padre decided to use the incident as the topic of his sermon that very Sunday. He praised the beautiful customs of Tlayacapán where all felt free to walk wherever they wished as long as they did no harm, but now, he said, there were new people in town, perhaps even dangerous people, who did not know or respect their ways, and so it was necessary to watch one's step.

"I had no idea who they were," Padre Chavo said later, "but I was very much a left-wing priest then, so I was not unhappy to have the opportunity to attack some rich people, which I assumed they were. In my text Ana Maria and Jean came out as the villains." The word about the sermon traveled fast, and as soon as Ezequiel heard about Chavo's criticisms, he told his *madrina*. She immediately went charging down to see the priest. Who did he think he was, and what did he mean by saying those terrible things about her in church, she wanted to know. Chavo was astounded. He said later, "I had never seen a woman like her. She was very angry, but I must say, I liked her for being so plucky and for saying exactly what she thought."

Unlikely as it would seem from that meeting, the priest and *la gringa* became friends. "While I explained the customs here, *she* explained that her husband was a mathematician and needed silence in order to work and that she was just trying to keep him from being disturbed." It was not until later that Chavo learned van Heijenoort had been Trotsky's secretary. "If I had known that," he said, "I would have gone to see him right away. I thought he was just another bourgeois *gringo,* and I even thought Ana Maria was an American because everybody called her *la gringa.*" Still later, the padre would change his own political outlook, but at that time he was very disappointed that Van was no longer a Trotskyite. "I was very much the revolutionary, anarchist, Maoist, Christian *fanático,*" he said. "I believed in all the *ismos,* and I thought it was horrible that he gave up the revolution for mathematics. Now I appreciate it, but then ... no."

There were other customs that made for tension between the locals and the outsiders. The *barrio* closest to the new house had a reputation for having more fiestas than any other in the village, and every time there was a party there were fireworks, orchestras, and bells. Anne-Marie would tell the revelers to stop making noise, but her orders and complaints had no effect at all except to create bad feelings. She would then go to Padre Chavo and vent her frustration, and he would listen. "She was anguished about many things," he said later. "She had a violent character, but we got along well. I understood her and I liked her. I thought of her as a good friend."

During the festival of Carnival, the van Heijenoorts invited the whole town to come to their house. Again, Padre Chavo was amazed:

> It seemed so strange to invite everyone to come and make a fiesta after the way they behaved earlier and after they complained so much about the fireworks, but the town orchestra and the whole village went and sang and danced in their house and she gave

everybody water and then they left and there wasn't a thing out of place.

Not long after Anne-Marie and Van's house was finished, they were told that the road which came to an end just past their property was going to be connected to a new road to Xochomilco which would link up with a highway to Mexico City. Although it would be a quarter of a mile away from the house itself, that was impossibly close as far as Van was concerned. Their friends, the Gollays, found great anxiety about this projected new road when they came to visit. While they could not understand the fuss about a little road that wasn't even built and wasn't going to be all that close, Van and Anne-Marie worried the subject to death, and she began to talk to Chavo almost daily about the road and also about her husband. Chavo said later that he began to think that

> This woman was kind of crazy. She would talk about what would or would not happen with the road and her house and her husband; and I, with my Christian obsession to save people, I would talk about what I wanted to do. But mostly *she* talked, and I listened and sympathized. I felt she came after me because I was like an empty pot she could piss in; she was extremely troubled about her situation, but at the same time, I was impressed even more by her strength and the sureness she had about herself. She was *una mujer muy macho* [a very macho woman].

It was Padre Chavo's impression that Ana Maria was "her husband's faithful servant," that she "adored her husband" and for that reason she became violent when anyone or anything bothered him or disturbed his work. But there was more to her anguish. By the late 1970s, Van had made it clear that he would not stay in Tlayacapán or Mexico City, and the question of where they would live was shifting to a question, as yet unspoken, of whether it was possible for them to live together anywhere.

Until he retired from Brandeis in 1977, van Heijenoort had always been on the move, traveling in his regular orbit, partly out of necessity, and partly as a matter of habit: France in the summer and usually at least once more in any given year; Brandeis and Cambridge in the fall; Mexico in the winter; and Brandeis again in the spring. During this time he was working on at least three projects: he was heavily involved in preparing the Trotsky Archive for its official opening in 1980, and as a spinoff of these efforts, he was completing *With Trotsky in Exile,* his own short book about his years in politics, and, finally, on the philosophical plane (as he would say) he was writing and compiling a series of papers he had written since 1967 that were concerned "with the place of logic in the body of human knowledge." These articles were among the essays that would be collected in the elegant volume, *Selected Essays,* published by Bibliopolis in 1985.

Being in constant motion did not in any way hamper van Heijenoort's ability to work. It is the reverse that appears to have been troublesome, for after he left Brandeis and told everyone, including himself, that he was moving to Mexico for good and would be spending most of his time there, he began to feel intellectually stifled. Not that he put it exactly that way. He always stated such problems first in terms of his physical well-being. In high altitudes he couldn't get his breath; in smoggy cities he couldn't breathe; in noisy places he couldn't think; in cold places he could never get warm. He took all the vitamins and minerals he could conceive of swallowing ("I take the whole alphabet," he would say). He was an avant-garde fiber eater, a cod-liver oil swallower, and abstemious of coffee, alcohol, and tobacco. He was attentive in the extreme to his intake and output, but in Mexico, it didn't help. Every close friend affectionately categorized him as a hypochondriac and a crank, but he thought he was just taking normal precautions.

Although the "noisy" new road was the scapegoat for his decision to leave Tlayacapán, the entire mountainous state of Morelos was also crossed off the list of possible places because van Heijenoort felt that even the 5,000-foot altitude was higher than he wanted to be, and probably detrimental to his health. Burdened with a set of spoken and unspoken conditions, he and Anne-Marie began to search elsewhere for the "absolutely right" place to live. Anne-Marie said that she would go anywhere Van wished, but, after the fact, there are those who were convinced that she didn't mean it. According to Laure van Heijenoort, in 1986, in a moment of terrible sadness, Anne-Marie said, "What would I do in the States? I have no family, no friends, no job. I would never leave Mexico." However, at issue in the late 1970s was the fact that van Heijenoort himself did not yet want to leave Mexico, or, if he did, he was not prepared to set up house with his wife in the United States. In contradiction to Laure's account of 1986, Anne-Marie once said that after Tlayacapán, there was a point when she came to the United States with her son, prepared to stay; and it was Van who wavered and would not make a clear decision about where to settle. According to Anne-Marie, in the end it was her husband who thought they ought to go back to Mexico and live there for a few more years. The only problem was to find the right place: warm, at sea level, quiet, picturesque, with simple people leading simple lives, close to an airport, with no smog or pollution.

Between more trips to the United States and Europe, they continued to scour Mexico looking for their part-time Shangri-la. It was at this point that it became clear that their marriage was in very serious trouble. The honeymoon thrill of being in constant motion was definitely over. In 1984, in an interview for this book, Anne-Marie described her feelings about that period:

> I couldn't go on like that. I had to take care of my son. I had my work to do. I loved Van very much but I realized that he would only do what *he* had to do, whatever it happened to be at that particular moment. He would only think about the things *he* was committed to. He wasn't considering what *I* needed. I felt insecure all the time because I didn't have any place where I could say, this is my house; this is my home. I would have gone to live with him in the United States, China, any place. I needed to have a point of reference, but he wouldn't say which was the place we should live. He wouldn't say, "this is it." Oh yes, he would pick a place and we would go there for a while, but we would never stay. For instance we made a trip to the Tarahuamara area [in the state of Chihuahua], and he said, this would be a nice place to live, so in my imagination I was already thinking where I would build a house and where I would put the rooms and where I would work—because women are like that—and then, he changed his

> mind for some reason or other. I could never make plans, and that bothered me very much. I was exceptionally happy with him except that thing about the house was always bothering me and it got worse and worse. I would walk on the streets where they sell things; I wanted to buy things, but I couldn't even look any more because I didn't have a house. I would ask him, "Where are we going to live? Where are we going to have a house?"
>
> And then he said to me, "I have no solution." For two years he said that: "I have no solution." Well that really upset me very much because all those years I was always thinking, maybe one day he is going to tell me, "This is it, this is the place," so when he said, "I have no solution," it was as if it was a mathematical solution he was thinking about, whereas I had been thinking of a human solution.

By 1980 they had reached an impasse and Van did not go to Mexico that year or the next, although he made no active decision to divorce or separate. He told some of his friends that it was finished with Anne-Marie, but he never said so directly to her. Anne-Marie, on the other hand, was direct. According to Van, she divorced him without so much as a by-your-leave.

Three years later, Anne-Marie explained her actions: "I don't like to have a messy desk, littered with papers. I like things neat and clean, and I like my life to be the same way: clear, even if it is harder." Although she claimed that she could never really be angry with Van, and that she never stopped loving him, in December 1981, she divorced him to marry Robert Duel, an American scuba diving instructor she had met in Cozumel.

Chapter Sixteen

# The End

1981–1986

We run carelessly to the precipice after we have put something before us to prevent us from seeing it.

BLAISE PASCAL, *pensée 183*

DURING THE COLDEST time of the year, Van Heijenoort had been in the habit of going from Cambridge to Mexico, but after his separation from Anne-Marie Zamora he felt he could no longer do that. Still, his urge for changing places was strong, especially because he considered the damp cold as one more enemy to his physical well-being. So, for the winter of 1981, he arranged to visit Stanford University, in large part because of the warm climate, but also because of its good libraries and congenial atmosphere. That he was depressed when he arrived was not immediately evident to those who did not know him well. His quiet, controlled manner seemed to be his normal style; he kept his problems to himself, had little or nothing to say about his personal life, and certainly made no mention of his wife.

On that first visit, he rented a sparsely-furnished room in a pleasant ranch-style house in a part of Menlo Park that had resisted gentrification and still retained some of its shacky rural charm. The house was on the street where Thorstein Veblen, the economist, had lived early in the century, and in the 1950s and 1960s the writer Ken Kesey and many of his friends had cavorted in the neighborhood—details which greatly pleased van Heijenoort. The owner, Patricia Griscom, was an attractive, soft-spoken woman in her mid-fifties, divorced, her children grown. She had several rooms which she regularly rented to students and occasionally to faculty who needed a place for a short-term stay. During what would be five years of comings and goings, Van stayed in three different rooms, but the one he liked best was that first small one, near the kitchen. He liked the comfort of kitchens, and although he had the run of the house, it was the only room he used besides his own. Every morning he invariably made himself a large bowl of *café au lait,* followed by a bountiful serving of cooked oatmeal. On evenings when he did not have a dinner engagement he also prepared himself a simple meal.

Originally, van Heijenoort's idea had been to stay only for the winter, but he chanced to come at the time when Solomon Feferman and others were formulating plans for a series of books containing the collected works, published and unpublished, of the great twentieth century logician, Kurt Gödel. Quite naturally, van Heijenoort, who had already done considerable scholarly work on Gödel (for his acclaimed *Source Book*), was consulted and then invited to be part of the editorial team of the "Gödel project." After some deliberation, he accepted. He had good professional reasons for doing so and what *he* thought were good personal reasons, which he kept to himself. Having made this decision he went back to Cambridge until the following summer.

In July 1982, before he left Cambridge for Stanford, van Heijenoort turned seventy. To celebrate the occasion, his friends and colleagues planned a surprise birthday party for him. Van Heijenoort, ever alert to nuance, got wind of it and berated Pierre Broué (who was visiting Harvard and in on the plans): "What kind of a fool do you take me for? Do you think I don't know what you all are planning?" With atypical anger, he told Broué he didn't want a celebration and he told him why. Until this point, he said, he had never thought about his age, he had felt no difference between being thirty-five and sixty-five; but now for the first time in his life he felt old, and he did not want to celebrate that. No doubt the outburst was provoked, in part, by the letter he had recently received from Anne-Marie in Mexico informing him that she had divorced him and married again. He was certain, he told Broué, that *he* would not be able to find somebody new.

Knowing something of Van's past, Broué was incredulous. He told him he was ridiculous, laughed and scoffed at the notion that at seventy, one crosses an arbitrary line that cuts off novelty, and said that, like it or not, there was going to be a party. In the end, Van laughed, too, and acquiesced, and on the occasion of the celebration, he seemed to enjoy the acclaim and congratulations of his friends and colleagues. But his general feeling of depression did not lift; it was a weight he brought with him to California when he returned to dig into the work of the Gödel project.

For van Heijenoort, living without a woman was out of the question. He was ascetic, he was nomadic, he liked to be alone, he liked to be quiet, he liked to be free to pick up and go any time he had the impulse, but he was a romantic who had to be "in love" or at the very least to have what he called "very warm feelings." So while he plunged into the heavy work of the first volume of Gödel's collected works with his unusual commitment to detail as well as content—worrying more than any other editor over the accuracy of translations, the aesthetics of the layout, and the use of commas—in his spare hours he agonized about what had gone wrong in his life. Because he was deeply upset, and perhaps because he felt time was running out—he often spoke of "coming to terms with my life"—he became far more communicative than he had been in the past. In letters and conversations, especially with women, he allowed himself to speak of his unhappiness. To his former student and friend, Anne Fagot-Largeaux, he wrote,

> *Je traverse une période de difficultés morales. L'apaisement et la sagesse que l'on dit venir avec l'age m'échappait toujours. Je ne vois devant moi que la nuit noir dans laquelle je m'enfonce.*
>
> [I am going through a period of depression. The peace and wisdom that is supposed to come with age eludes me. I see nothing ahead of me except the black night into which I sink.]

Later, when he saw Fagot and his other friends in Paris, he repeated what he had said to Pierre Broué, "If I were younger I would begin again, but now . . . ." Nevertheless, if not right then, soon, in spite of what he said, he did slowly begin again.

In the Menlo Park house, Van kept to himself for the first few months. Pat Griscom's first impression of him was of "a very dignified gentleman, who was incredibly charming, who, without seeming in any way, shape, or form unfriendly, was able to let you know that he was also a very private person."

By nature, Griscom was, in her words, "very accomodating" and more than willing to hold back and made sure that van Heijenoort had as much "space for himself as he might want." It took some time, then, before any kind of real friendship developed between Van and his landlady. The catalyst was a visit from van Heijenoort's daughter. With Laure present, Pat noted a marked change in Van's behavior: "He was so pleased that she was there; he was so proud of her. He walked into the kitchen and said, 'Don't you think she looks like me?' Until that moment, he had never made a personal comment to me about anything or anybody."

While Laure was there, the three of them had some meals together and went on one or two excursions, and after his daughter left, Van bought Hayden Herrera's biography of Frida Kahlo for Pat, as a small thank-you gift. He had presented that book to many of his friends because he thought it was good, because he had been an important source for the author, and because he played a cameo role in the text. It was a way of letting people know something important about him without having to tell it himself. Pat started reading the book and suddenly realized who her roomer was. The next day she knocked on his door and said, "Well, Monsieur Secretary, why didn't you tell me?"

The ice was broken, the atmosphere changed, and although van Heijenoort made his usual circuit to Cambridge, Paris, and other places (except Mexico), and kept his rooms on Kirkland Place in Cambridge, the house on Perry Lane became more like home. He and Pat went to movies together, occasionally to dinner, and sometimes on Sunday excursions to the beach. When Pierre Broué came to the Hoover Institution the following year to work on the Trotsky–Sedov papers, Van arranged for him to stay in Pat Griscom's house and there was more catalytic action. Pat was amazed at the way these two mature Frenchman carried on, like rowdy schoolboys. Broué teased Van, mercilessly, and Van gave back as good as he got. When he spoke French, van Heijenoort had a way of relaxing and dropping his guard that was a revelation to those who had only seen him in his proper, American version of himself. This was never more true than with Broué, for in addition to the general cultural background they had in common, they spoke an almost private language based upon their intimate knowledge of the Trotsky years.

The gaiety and good times were fine, the Gödel work was fine, the archival work with Broué was fine, but all that was only on the surface. Inside, where it mattered, van Heijenoort still felt unhappy and incomplete. The only solution he could imagine was to reunite with Anne-Marie because, he now realized, he "was very attached to her"—which was his way of saying he needed her. The problem before him was how to do it because, in the first place, she wasn't speaking to him or answering his letters, and, in the second place, she was married to someone else.

Two years earlier, when the ten-year "honeymoon" of Anne-Marie and Van had ground to a halt, ostensibly over their non-agreement about where to settle down, van Heijenoort had not been sorry about their separation and what appeared to be the end of their marriage. Even if his wife *had* agreed to live in the United States, as she claimed, it is not at all clear he would have wanted her to. But, when she actually divorced him and married someone else, he responded in a way familiar to the women who knew him well. Van's former wives and lovers had all labeled him jealous and possessive and unwilling to let go, even after he had initiated a separation. All it took was a serious relationship with another man to set off an emotional frenzy during which, with alternating doses of threat and charm, he would plead for a reconciliation. In the past, the tactic had worked, and although it was an old game, he was ready to try it again with Anne-Marie Zamora. This was the hidden "personal" reason for his having accepted the editorial position in California: it was easier to travel to and from Mexico, and also, Anne-Marie might be persuaded to come to Stanford.

Meanwhile, although he kept his part of the editorial work flowing and did many things to divert and entertain himself, he remained troubled. "The nights are the worst," he would say. "I can't sleep, or if I do fall asleep, I wake up at 3:00 in the morning, and that's it." Sleepless, he read and admired (to no one's suprise) the memoirs of Casanova and Madame de Staël, the histories of Thucydides, and every good novel recommended to him; and after that, "I replay my life and think of what I should have done." But when he said, "I should have had one wife, one home, and one work," it was a statement guaranteed to make his friends laugh and shake their heads.

Persistence was one of van Heijenoort's strong points, at least until it was overtaken by an unreasoning stubbornness. Never one to give up on what he thought was right or what he thought he needed, the ex-husband courted his ex-wife by letter and telephone until eventually she agreed to allow him to see her. Walking on eggs, he made a few quick trips to Mexico in the summer of 1984 and upon his return thought he was making progress. "Of course nothing is certain," he said, "Anything can happen. Nothing is excluded." But indeed the progress was so good that Anne-Marie agreed to visit him in November and there were hints that she might consider living in California, if and when they reestablished their marriage. As usual, van Heijenoort gave his friends very little information. "It is all very delicate," he would say.

Not many people at Stanford had occasion to see Van together with Anne-Marie during her brief stay in the fall of 1984, but those who did marveled at the transformation in him. Pat Griscom, who had not known anything about his ex-wife until he brought her to the Menlo Park house, later said, "He was so happy. He was just beaming." However, Pat had the impression that Ms. Zamora was somewhat less euphoric, and observed:

> She didn't smile, she didn't do anything. She just stood there in her little dressmaker dress. And when they left for Mexico, Van did not do his usual thing of kissing me on both cheeks, the French way. Her jealousy must have started at that time, or else he was worried about it, but I myself was oblivious of it.

As a suitor, the former husband was tender and attentive towards his ex-wife/wife-to-be. At dinner at a friend's house, he hovered over her, watching and encouraging her to eat. After the meal, he sat her in a large wicker chair and spread her long reddish-blonde hair out across the back of the chair and stood, admiring. All this he did without a trace of self-consciousness, but simply involved in presenting Anne-Marie in the best light possible. His conversation was of the goods "we bought and we will take back to Mexico." His use of the first person plural was as surprising as anything he did that evening since "I" was the pronoun he usually used, even when speaking of his activities with others. As for Anne-Marie, she was quiet, but not unfriendly, and by the end of the evening became more animated.

Although van Heijenoort continued to act as if nothing was certain, Anne-Marie was making more definite statements about her feelings and plans, saying,

> I think we will get married very soon. I love him very much. I never stopped loving him. It was just that he made me angry when he would never make a decision about where to live.

Another reason Anne-Marie gave for her change of mind was her son's impending marriage—a union she opposed. With a surprising lack of embarrassment she confessed that she did not want her son (who was in his mid-twenties) to marry and to leave her house, but that in spite of her wishes he was going to do both. "So now I need Van more than before," she said. "He is all I have left."

Anne-Marie never mentioned Robert Duel, to whom she was still married. It was as if he was not a factor in her decision. And, if van Heijenoort's friends were taken aback by her explanation of her change of heart, he was not. She needed him; he needed her. If their reasons were different, at least for the time being, his depression was lifting.

"So they went off into the sunset," Pat Griscom said later, "and I never expected to see him again." Van Heijenoort's careful strategy had apparently succeeded; Anne-Marie promised to move to some place on the coast, north or south of Acapulco, but in any event, away from the high altitude of Mexico City and the impossible noise and pollution that van Heijenoort could not tolerate. In December 1984, three years after Anne-Marie had divorced him, they were reconciled and remarried. There was no evident difficulty in regard to Ms. Zamora's interim marriage. She simply divorced Duel and re-married Van in one swift operation. Van Heijenoort told Solomon Feferman that he was moving to Mexico, but not to worry: he would continue his work on the Gödel project and make trips to Stanford as often as necessary to consult with him and the other editors. Some of their business could, of course, be done by mail or telephone.

After a honeymoon in Costa Rica, the couple began their search for "the right place to live." Then, in the immortal words of Yogi Berra, it was *déjà vu* all over again. As before, the search for Shangri-la was futile. Anyone but Van would have seen it coming. In the past, whatever the cause of the rupture, van Heijenoort's reconciliations had always ended with his realizing "it was too late," that it wasn't going to work. The mystery is how he could have deluded himself into thinking that this time things would be different. Françoise Ville, one of his closest friends, later suggested that perhaps Van was indeed feeling his age, and that the man who had been afraid of almost nothing was finally afraid of growing old in some corner, by himself.

There are conflicting opinions as to whether Anne-Marie ever believed that Van really became physically ill in high altitudes. However, if she thought the problem was psychosomatic and part of his general hypochondria, she was not the only one who believed that was the case. There is also ambiguity as to whether she was serious about being willing to live outside of Mexico City. But in the end, all the talk of "place" seems to have been a metaphor for finding a way of living with each other in the long term. As Anne Fagot-Largeaux put it, "They could never find the right place because his *place* was not with her."

* * *

During the six months that Van and Anne-Marie were together again in Mexico, Olivia Gall, a graduate student in political science, came to call on Professor van Heijenoort for advice about her thesis work, which analyzed Leon Trotsky's influence on Mexican politics. Gall had been Pierre Broué's student at the University of Grenoble and had been introduced to van

Heijenoort early in her student career when she had come with Professor Broué to study the Trotsky Archive at Harvard. Subsequently she had interviewed Van in Geneva and at Stanford. On all these occasions he was willing—as always with students—to describe the intricate details and time sequence of a particular situation, or to paint a broad overview, or ready to help with references and suggestions. As Gall's knowledge of her subject grew, so did her appreciation of van Heijenoort's expertise.

Gall had been present at Stanford in 1984 while van Heijenoort and Broué were deciphering the newly found Trotsky-Sedov letters. At that point she had begun to understand the true depth of his knowledge about the communist, socialist, and opposition movements of the left. Seeing Van and Broué working together on archives, she had a new appreciation for the unique value of his work on the archives: he was indeed literally irreplaceable. Above all, she respected his "extremely critical but not reactionary point of view regarding Trotsky as an individual and as a political figure." And finally, Olivia admired Van's personal style. "I was impressed by how well he took care of himself," she said later, "—the healthy food, the long walks."

When she saw him a year later in Mexico City, Gall found van Heijenoort quite changed physically. In California she had thought he was "in great shape," but now he was pale, complained of the flu, amoebas, and other problems. "He was tense, his manner was formal, and Anne-Marie's big house did not seem like the right place for him. I thought he was much more himself in the simplicity of Pat Griscom's house in Menlo Park." Olivia had come to discuss the form and writing style of her dissertation as well as matters of French usage (her Ph.D degree would be from the University of Grenoble); it was understood that van Heijenoort would not want to interfere with Broué's role as advisor on content. Even so, Gall was taken aback at his formality and restraint. "He said only what was required, nothing more, and no matter how hard I tried I was not able to break through the iceberg that had installed itself in him."

She had a mixed reaction to Anne-Marie, whom she found "more sexy than beautiful, quite nervous, a bit odd, direct, somewhat cold yet with a certain sweetness that was hard to define."

Gall's mother lived in Mexico City; and wanting to reciprocate Van's past kindness to her in the United States, Olivia invited the van Heijenoorts to a dinner party at her mother's house. That evening, she later said, "I discovered yet another side of Van":

> He was a man crazy in love, like an adolescent, so physically, sensually, and sexually attracted by his wife that not only did nothing else seem really important to him, but also his vision of the world and his political opinions—all that refinement and critical spirit—were subordinated to those of his wife. We spoke of Nicaragua, where Anne-Marie had just been, and their point of view was so extremely reactionary that I could not get over it—not that my family and I were unconditionally in favor of the Sandinistas. The atmosphere became very tense and I was happy when they took their leave. As I accompanied them to their car, they walked in front of me holding hands and looking at the starry sky, and again Van wore the smile of a hypnotized adolescent.

No doubt there were many undercurrents in the atmosphere that Gall could not have appreciated—for instance, van Heijenoort's disinclination to differ with his wife in public. Also, she was not the first to have been surprised that his political views, when he expressed them, were far to the right of most of the students and scholars writing about Trotsky. Finally, though only a few people were privy to the view of "Van in love," those who were, all expressed similar astonishment at the metamorphosis he underwent in that state of mind. A friend from the 1960s said that, by chance, she once saw him meeting a woman on Fifth Avenue in New York: "He swept her off her feet, into his arms and, embracing her, swirled around and around. It was the most romantic thing I have ever seen in my life."

A few months after the dinner party, van Heijenoort telephoned Olivia Gall, told her that he was leaving Mexico, and asked her to come and see him at the Hotel Reforma where he was staying. They talked about her work: he insisted that even if she was having difficulties she absolutely had to finish. He offered to help by correspondence, and he promised to be a member of her jury when she defended her thesis. Then, very discreetly, he told her of his troubles with Anne-Marie, saying, "It's not working; it's not working. It's hard to admit but it's been going on like this for years."

In the spring of 1985, while van Heijenoort was still acting as if Mexico was his main place of residence, he made business trips to Stanford and to Cambridge, Massachusetts. When he was in Cambridge, coincidentally, Pat Griscom happened to be visiting relatives there. Van was delighted to see her, and immediately invited her to lunch at the Harvard Faculty Club. He now thought of Pat as a friend he could confide in, and although Pat did not have the premonition that Anne-Marie had had fifteen years earlier when she knew "something would happen if I accepted the lunch," Griscom, too, acknowledged the meal at Harvard as a prescient moment:

> Something clicked. There was a different kind of chemistry between us, nothing explicit, but something. We both realized we were having a very good time.

Van Heijenoort went back to Anne-Marie in Mexico and Pat went back to California. A few months later, Pierre Broué came to Stanford to continue the work on the recently resurrected Nicolaevsky papers. Van had promised to come at the same time, to help him decipher and identify names, contexts, and handwriting. The *Gödel* editors were also anticipating van Heijenoort's presence at their summer editorial meetings, but everyone thought he would stay for only a limited period. However, on his arrival, Van announced that his return would again be "permanent." In little more than half a year, he had again concluded that "Mexico is impossible." All his old symptoms had returned; Anne-Marie did not understand his problems; and Anne-Marie had not found "the place." There was one new item added to the list: he complained that his wife was jealous of his work and was making it impossible for him to do anything. He had no choice; he had to leave.

Saying *"c'est fini,"* van Heijenoort moved back into the house in Menlo Park, this time with all his worldly goods. "I had no idea he had so much stuff," Pat Griscom later said, "suits and sweaters, things that he had left in Mexico since the seventies. I had thought he was a man of few possessions." He resumed his routine of early breakfast, the walk to Stanford, the undistracted engagement in his work, and, in almost no time, it dawned on him that his life was fine, that he could live very happily in Pat's house.

Suddenly everything was possible, even the kind of new relationship which two years earlier he had said was impossible. One evening, within a few weeks of his return, he surprised Pat by asking her if she would "come to him." Pierre Broué, who was living in the house for the summer, saw them come out of Van's room one morning, and from the way they looked at one another, there was no question in his mind that they were lovers.

For a few brief months, just before the end of his life, van Heijenoort actually experienced the tranquility and quiet love he had been seeking. He called Pat Griscom's home "a haven." He had, as Griscom said,

> a pleasant woman who grew to care about him very much, and a place where he could go on with his work. He became very domestic, putting meals together and liking to be praised for his cooking. He was like a kid. If you asked him to tell you how he did something, he had this little smile as he told all the details. He also taught me how to use his computer, and we read Turgenev together which he liked because the settings were so peaceful.

In November 1985 Van made one of his regular trips to France, and saw his usual crowd including his old friend, Anne Fagot-Largeault. They went to dinner, as they often did, and, although many aspects of their evening together were like other times, Dr. Fagot-Largeault remarked upon a difference in his mood, a sharp contrast to the depressed tone in his letters of the previous years:

> He was in very good spirits that night. When we met he kissed me in the French way [a kiss on both cheeks] for the first and only time. As usual, we talked about a lot of things: literature, academic life, logic, the Gödel project, and more personal matters. But then he told me that he was through with his wife. "*C'est fini*," he said.
>
> After dinner, we walked back from Les Halles and near the Tour Saint-Jacques we saw a beautiful *manège* [carousel], an old fashioned one, with wooden horses. He stopped and stared at

it and said, "Let's go for a ride on the *manège*." I looked at him and I saw that he was serious. So we did it. And he was happy, *joyeux!*, on that merry-go-round horse. It was wonderful. I felt a kind of complicity but I must say, I was very surprised.

Typically, Van did not reveal the source of his high spirits; he did not tell Anne Fagot that after it was "all over" with Anne-Marie, he had begun a new relationship with someone else. Perhaps, for all his assertions, he knew the emotional strands between Anne-Marie and himself could not be untangled that easily. Even before his visit to France there had been some signs of what was to come.

On September 19 and 21, 1985, the center of Mexico City was struck by two cataclysmic earthquakes, the first 8.1 and the second 7.5 on the Richter scale. The devastation was staggering, the unofficial low count was 20,000 dead. Van Heijenoort immediately tried to telephone Anne-Marie, but in the ensuing chaos it was impossible. His instinct was to jump on the next airplane but the Mexico City airport was closed. Although he was reassured to learn that there was almost no damage in the Lomas de Chapultepec area where Anne-Marie lived, he did not rest until he was able to reach her and hear that she was not harmed. With a hint of pride he reported to his friends that she was active in some kind of quake relief. Then as soon as he could, he flew to Mexico City for a short stay to reassure himself further.

Three months later, despite plans he and Pat had made to be together, Van went to Mexico to spend the Christmas and New Year's holiday with Anne-Marie. He did not mention his new plan until two or three days before he left, and then, he simply said he would be going away, without saying where. Naturally, Pat was hurt and angry, and when Van came back in January, there was a definite chill in the atmosphere in the house on Perry Lane. Still close-mouthed about exactly where he had gone and what he had done, Van did acknowledge that he had gone to Mexico and added, "It was a mistake." The "mistake" (one might safely guess) was that he and Anne-Marie had, once more, gone in search of a place to live—it had become a way of life between them—and had again failed to settle on a choice.

Although Van was back in Menlo Park, it was soon obvious that Anne-Marie had an enormous pull on him. He was obsessed with her and deeply conflicted about what to do. At times he told Pat that he was sure things would work out between the two of them, but Pat had many misgivings about their situation. Meanwhile, if Van was confused, Anne-Marie was not. She knew she was not willing to accept her husband's decision to leave. For her, it was not finished and would never be. She had taken him back, she had set her sights on a life together, and his re-rejection was a loss she could not accept.

In a desperate attempt to hold on, Anne-Marie launched a relentless barrage of telephone calls at all hours of the day and night, first to the house in Menlo Park, until Pat told her she had to stop, and thereafter to van Heijenoort's office at Stanford. She threatened to kill herself and to kill him, and she began to give deadlines.

Clearly Anne-Marie was unbalanced. In retrospect, a disturbing question arises: Why was it that no one at any point took measures to protect her from herself or to urge her to seek professional help? She had had earlier deep depressions and emotional crises. Did no one take her seriously? Did people think she was just being dramatic? Had she cried wolf too often?

On some level, van Heijenoort certainly believed her. To close friends he reported, "Anne-Marie says she is going to kill herself, and she wants to kill me too," but he said it in his everyday, understated tone. Only if they took the stance that Anne-Marie was merely using a figure of speech would he become agitated and say, "No. She means it! She is capable of doing it!" If one then said, she is disturbed, she needs help, perhaps she should be in a hospital for her safety and yours, Van would throw up his hands and say that was impossible, that she wouldn't go, and that if she were to be committed there was no institution that could hold her because she would either talk or buy or threaten her way out. For many months just before and after their remarriage in 1984, Van and Anne-Marie had gone in search of a cure for her mood swings which, they said, were caused by a "chemical imbalance" that had never been properly adjusted after a hysterectomy she had had some years back. They consulted doctors at famous clinics and hospitals in California, in Massachusetts and in France, and always, they reported, the doctors could not prescribe the right mix of medication to set the hormonal levels right.

Van Heijenoort had little confidence in the ability of conventional medicine to effect cures and in general held a low opinion of doctors, but that never kept him from seeking their opinions. He was perenially searching for a "good man" who would hit upon exactly the right medication. At the same time, he was a health nut and a devotee of over-the-counter concoctions; he had his own prescriptions of large doses of vitamins and minerals and patent medicines (DSMO, for example) which he took himself and urged upon others. But in Anne-Marie's case he had to acknowledge that *his* cures didn't work either. As for psychoanalysis or psychotherapy or counseling as a means of dealing with the situation, it appears not to have been considered.

Either through an unwillingness to acknowledge the depth and character of Anne-Marie's distress and the role he played in it, or simply by habit, van Heijenoort persisted in thinking that "reason" could solve their problems. "Why can't two *reasonable* people sit down and come to a *reasonable* solution?," he would say with imploring innocence, oblivious to his snarled history of vacillation and wandering.

The same January in which Anne-Marie Zamora was using the telephone lines to hold on to her husband, Dr. Giovanna Corsi, a logician from Florence, came to spend the winter quarter at Stanford. Because there were many visitors and work space was tight, van Heijenoort graciously agreed to share his office with her. Until that point, Corsi had known van Heijenoort only by reputation as "a great man, a distant figure," but now, inadvertently, she became privy to his personal problems because she happened to be present at the beginning, and sometimes at the end, of some very long and obviously difficult conversations.

It was an embarrassing situation for Dr. Corsi, a discreet and tactful person; she did not know quite what to do when a call came. As soon as she realized the matter was something deeply personal, she would leave the office and go to the library. A half-hour later, she would return to find the agitated conversation still going on, but if she started to leave again, van Heijenoort would make motions indicating that she should stay. Sensing that he was concerned about driving her from the room, she returned to her desk and tried to pretend she was engrossed in her work.

According to Corsi, the telephone conversations would go on for an hour with Van Heijenoort carrying on as if Corsi were not there, pleading, scolding, and arguing. (She confessed that in a certain way, before the final tragedy was played out, she was rather pleased to discover that the dignified Professor van Heijenoort, at his age, was in the throes of such a passionate relationship—a man of 73, still feeling and provoking that kind of intensity and anguish.) When van Heijenoort was not in the office, Dr. Corsi answered the telephone and was subjected to persistent and insistent questions about where he was, when he would return, and at what other telephone numbers he might be reached.

In February, van Heijenoort was scheduled to go to France. Two weeks before his departure Anne-Marie began calling three and four times a day. At this, van Heijenoort lost all patience, exploded in anger, and sometimes hung up. Still the calls did not stop. Seemingly at his wit's end, he then changed his tone and his tactics and suggested to Anne-Marie that she come to Paris with him, and, in response to her questions, gave details of his itinerary and flight. It was Ms. Corsi's impression that he did so only in the hope that she would stop harassing him with phone calls and that the last thing he wanted was for Anne-Marie to join him.

The specific impetus for van Heijenoort's trip abroad was Olivia Gall's Ph.D. oral examination at the University of Grenoble. There he would make good his promise to be a member of her jury and have the satisfaction of seeing the work he had encouraged completed. The prospect had pleased him very much, but the solo trip, as he had envisaged it, was not to be. Anne-Marie called his bluff; taking him at his word, she was there, at the airport in New York, ticket in hand, ready for the flight to Paris. How surprised was he, truly? Knowing Anne-

Marie as he did, it is hard to imagine that Van was completely taken aback. He, after all, gave her the details of his itinerary.

From New York, they flew to Paris—together. In the close quarters of the airplane cabin, Van was forced to acknowledge the full measure of Anne-Marie's despair and the lengths to which she was willing to go to dispel or display it. And he saw that his half-business, half-pleasure trip had the potential to turn into a nightmare from which it would indeed be difficult to awake.

Van Heijenoort's original plan had been to stay with his daughter Laure, who happened to be spending some months in Paris. She was completely mystified when Van arrived at her apartment accompanied by Anne-Marie. In all her recent communications with her father, she had been hearing nothing so much as descriptions of the problems and tensions between him and Anne-Marie and of the impossibility of his ever living with her. Van offered no explanation; after greeting his daughter, he simply said that since Anne-Marie was with him, they would go to a hotel. Laure noticed that Anne-Marie was unusually subdued and that they both looked terrible, but her first thoughts were only that they were tired because of the long flight.

The exhausted couple went to the Hotel de la Sorbonne, across from the Lycée Saint-Louis, Van's old neighborhood. To control her agitation, Anne-Marie was taking large doses of medication, including fifteen to twenty tranquilizers a day, by van Heijenoort's count. In his opinion, four should have been more than enough to put any normal person to sleep. But for most of that week, Anne-Marie was unable to sleep, and neither was Van. By two in the morning she would be in a high-pitched state of hysteria, crying and talking, telling him he did not love her, and accusing him of being Olivia Gall's lover. She would ask him what she should do, then answer her own question by declaring that she should kill herself and him too. During this crisis van Heijenoort called almost no one in his circle of friends to announce his arrival or make dates to see them. He was afraid to leave Anne-Marie alone, and he certainly was not going to take her out into company. He did call his son to say he had arrived, and he let his daughter know some, but not all, of what was going on. He told her that he had not been able to sleep because Anne-Marie was disturbed.

Meanwhile, from Grenoble, Pierre Broué, worried that he had not heard from his punctilious colleague who usually called the moment he was in Paris, telephoned Van's son and learned, to his dismay, that Van had arrived with Anne-Marie. Broué then made many unsuccessful attempts to reach his friend. On the other hand, van Heijenoort did call Olivia Gall, who had come to Paris well before her scheduled Ph.D. orals. He asked her to bring her thesis to him at the Place de la Sorbonne, telling her that Anne-Marie was with him and that she was not well. "It's an old problem," he explained, "a psychiatric problem. She is in a depressed state and I cannot leave her." Gall, too, was surprised to hear that Anne-Marie was present. Just a few weeks earlier Broué had told her, with some satisfaction, that it seemed that Van and Anne-Marie had separated definitively. Never one to conceal his sentiments, Broué went on to say he was happy about it because he had thought their relationship was destructive, and finally, speaking metaphorically, he said, "If he had stayed with her, she would have ended by killing him."

When Gall brought her thesis to the hotel she was startled by her mentor's appearance:

> As he came downstairs to meet me, he looked just like the Van who had come down the stairs at Anne-Marie's house in Mexico. He was pale, gray, and faded; his voice was so weak that I wondered if *he* wasn't the sick one. He took the thesis and promised to read it as soon as possible, "as soon as Anne-Marie gets a little better," he said. I wanted to ask him more, to ask if I couldn't help somehow, but his cold formal side took over and it was impossible to penetrate that.

Towards the end of their first terrible week in Paris, Van and Anne-Marie sought the help and advice of Dr. Lea Beaussier, a friend of Van's since his student days. Later, describing the character of their consultation, Dr. Beaussier made it a point to state that she was definitely not a "counselor in psychodrama or a psychiatrist." Her role, as she saw it, was that of an understanding friend with medical knowledge. "In their state of *crise conjugal,* Anne-Marie and Van came to see me twice and I did what I could to help, as a friend." Dr. Beaussier's professional concern had a temporary calming effect on Anne-Marie and Van; her recommendations were that Anne-Marie return to Mexico as soon as possible, to be hospitalized or at least placed under doctor's care. Van Heijenoort was more than willing to have his wife follow this advice, but he did not think she should go alone. On the other hand, since he was sure he did not want to accompany her, he tried to think of someone who might do so. What Anne-Marie thought about all this is not known.

On Sunday, Van's half-sister, Paulette Corpion, invited all the van Heijenoorts to dinner, and everyone, including Anne-Marie and Van, gathered in Creil at the house where *Maman* Hélène had lived into her nineties. Laure and her companion, Bill Walker, were there as well as Jean, *fils,* and his wife, Yveline. Paulette knew nothing of the current situation between Jean (as the family always called him) and Anne-Marie but it was obvious to her that something was very wrong. She was accustomed to her brother's withholding silence, but she found the usually talkative Anne-Marie's "*silence hermétique*" very strange and disturbing. Anne-Marie hardly touched her food, and this, at one of Paulette's superb Sunday dinners, was a sign of trouble.

In the middle of the meal, Anne-Marie excused herself, saying she needed to rest, and went to a bedroom to lie down. Laure followed to see if there was something she could do. To Laure, Anne-Marie lamented that Van no longer loved her, that he was leaving her, that he would not live in Mexico, and that *she* could not live anywhere else—alone, without family and friends.

The following week, three days ahead of the oral examination, van Heijenoort and Anne-Marie took the train to Grenoble. Broué, who had not seen Anne-Marie for seven years, found her appearance frightening: "She looked drugged and *hallucinante.*" In the few minutes that he had alone with Van, he expressed his concern. Van responded with a full account of the nightmarish week of sleepless nights, jealous accusations, hysterical weeping, and threats of suicide and murder. Later, Broué recounted the rest of their conversation:

> I became even more afraid and told Van that I thought Anne-Marie was really capable of killing him, to which he replied, "Oh yes, she is capable of killing not only me but also any one of my friends–you, for example!" He was totally exhausted and beaten and kept repeating, "It's beyond me, it's beyond human force. There is nothing I can do."

Broué had little sympathy for Anne-Marie. He was visibly angry that she had come to France and thought it was incredible that Van was not able to leave her once and for all. In contrast to others, Broué thought her behavior was *deliberately* outrageous. As evidence, he later reported that during her stay in Grenoble Anne-Marie had called his wife, whom she had never met, and asked her the best way to commit suicide. Should she throw herself off the *téléphérique*?, she asked. Even though he had described her as *hallucinante*, Broué could not excuse such talk.

With considerably more than the usual anxiety that surrounds Ph.D. orals, Olivia Gall worried about whether Anne-Marie would attend and what she might do to disrupt the exam. Broué, with the intention of cautioning his student, had told her all that Van had told him, including Anne-Marie's accusations that Van was having a love affair with her. This, Olivia said, she found impossible to believe. But in spite of the high tension, the examination proceeded and concluded without untoward incident. Anne-Marie was indeed present; she looked distraught, but, like everyone else in the room, she sat quietly and listened attentively.

*Jean van Heijenoort at Olivia Gall's Ph.D. orals. Grenoble, 1986.*

*Pierre Broué and Van listening to Anne-Marie sing at Olivia's Ph.D. party. Grenoble, 1986.*

There was, however, some unexpected drama. Stepping out of the role of examiner, Professor van Heijenoort, the historian–archivist rose to the occasion and gave a bravura performance. As Gall reported:

> He was more than just a member of the panel, asking questions. He was an eyewitness to the events we were talking about, a witness who surprised us with anecdotes about Trotsky that he seemed to have prepared and saved just for this moment. As he told these stories, his manner changed completely. Looking at him and listening to him, I forgot about how he looked at the hotel in Paris, and I forgot that Anne-Marie was in the room. He was himself—or rather that aspect of him that *I* called "himself"—because, in reality, as all the evidence shows, his attachment to Anne-Marie was also a part of him that could never be undone.

After the presentation, there was a small party in honor of Olivia which Van and Anne-Marie attended for a short while. Again he became very tense and never moved more than a meter's distance away from his wife. At a certain point, she asked to borrow a guitar, and she sang some Mexican songs in a voice Pierre Broué later described as "absolutely lacerating" and Gall said "made your hair stand on end." Towards the end of her performance, there was a moment when she became confused and seemed to forget the words of the song she was singing and ended by finishing a different one. Then, she turned to her husband and said, "Van, stay with me, stay here with me. Don't go away."

Before leaving the party, Anne-Marie took Pierre Broué aside and said that since he was Van's friend he ought to know the truth. "I love Van," she said to Broué, "but he doesn't love me any more." Broué answered, "I cannot talk about that, as much out of ignorance as principle." She persisted, saying, "Tell me something that will help me." Then Broué said that he had seen Van very sad when they had been separated and very happy when she called him back, but that he could assure her that it was really physically impossible for him to live in the altitude of Mexico City; and finally he said that he was honored to have known and been close to Van, and she answered, "Me too."

The following day Van and Anne-Marie took the train back to Paris. By chance, at the station in Grenoble, they met Olivia Gall. Once more, she thanked van Heijenoort for his help and his presence and ask what his plans were. He told her that he would be returning to Stanford the following week, then going to Harvard in March, and that he was not planning to be in Mexico in the near future.

It was cold and raining in Paris when Anne-Marie and van Heijenoort arrived at the Gare de Lyon late in the afternoon. The rush hour had begun and the station was jammed with commuters. As they walked through the station toward the main door to find a taxi, Anne-Marie disappeared. In his explanation to Laure, Van later said that at a certain point, he looked around and she was gone—vanished in the crowd. He retraced his steps back to the train platform, searched the waiting room, and then gave up. Stunned, exhausted, feeling himself on the verge of sickness, he concluded that Anne-Marie had deliberately walked away. Instead of going to a hotel, as planned, he took a taxi to his daughter's apartment and went to bed.

Was he telling the whole story of what transpired between them? Did Anne-Marie really vanish without a word? No one knows. If Van was worried about Anne-Marie's disappearance, he was also relieved. He had reached a point of numbness, beyond caring. He developed a terrible cold which kept him in bed for several days, and he focused his attention on healing himself by employing his usual regimen of heavy doses of vitamins, minerals, and oatmeal. According to Laure, since he heard nothing further from or about Anne-Marie, and on the theory that no news was good news, they presumed she had acted on Dr. Beaussier's advice and gone back to Mexico, albeit alone. If Van called Mexico City to confirm this, he did not tell anyone about it.

In the remaining days of van Heijenoort's stay, he and his daughter went to a dinner party given in his honor at the apartment of Philippe de Rouilhan, a young colleague and friend. Van Heijenoort had taken a great interest in de Rouilhan's work and also in his personal life. De Rouilhan realized that he was being treated with an unusual, almost fatherly affection, and he was flattered and deeply touched by Van's concern. He had planned the party with great care, giving much thought to the guest list and the menu. Philippe and everyone involved anticipated the evening with pleasure.

The timing could not have been worse, but neither the host nor the other guests, with the exception of Laure, were aware of the emotional trials Van had been through during the past weeks. De Rouilhan, a man with considerable *savoir-faire*, had no idea what was happening as each topic of conversation eagerly directed towards van Heijenoort fell into the void of heavy silence in which he had buried himself. "It was a disaster. More like a funeral than a party," Philippe said later. "I am from the *Midi*; I know how to be charming, but there was absolutely nothing I could do." It was not until the following year that the bewildered host understood the background for his friend's behavior that evening.

Van Heijenoort called upon one last person in Paris, Françoise Ville, a good friend about whom he once said, "We could stay close for so long because we were never lovers." Never lovers, but very empathetic, they understood each other well. According to Ville,

> He came to see me alone, he looked awful, like he was about to collapse. He told me about the violent scenes with Anne-Marie, he made these clumsy gestures and said, "I can't stand it, I can't stand it, I can't take any more." I had the impression that there was a terrible crisis at that moment. When he left, we said, goodbye, goodbye, see you soon, because he was supposed to come back in a month. It was the last time I saw him.

* * *

Van Heijenoort returned to California on March 5, 1986. According to Pat Griscom, who picked him up at the airport and brought him home, he was haggard and exhausted. Only to her did he recount the awful details of what had happened in Paris. To others who asked about his trip, he said, "Paris was nasty; miserably wet and cold. I caught a cold; I didn't have too good a time." He made no mention of the fact that Anne-Marie had been with him and then disappeared, and consequently, he did not speak of their violent quarrels, or of her drugged, depressed state. Except to Pat, it appeared that his mood lifted as soon as his cold was gone. He resumed his regular pattern of work: the Gödel project in the morning, at his office, and the Nicolaevsky papers in the afternoon, at the Hoover Institution. He was also reading Thucydides and thinking about writing "something not too long" comparing Trotsky and Thucydides and their perspectives on history. He had long thought of writing an intellectual history of Trotsky, but the comparison with Thucydides was a new idea.

Dr. Giovanna Corsi was still at Stanford in March and still at the other desk in van Heijenoort's office. She, too, was completely unaware that the woman who always telephoned had actually gone to France with him, and she knew nothing about what had happened there, but she *was* aware of a difference in van Heijenoort. When the telephone calls began again, as they did almost as soon as he returned, he no longer sought to end the conversation quickly. He did not get angry: he listened, and he was sympathetic. He even said he would go to Mexico as soon as some dental work was finished. Nevertheless, it was Corsi's impression that

> it was over between them. I thought the problem was that she (Anne-Marie) was not ready to accept that. She didn't know what to do; she didn't know how to handle the breaking up of their relationship. She just kept on calling, all the time. If van Heijenoort was not there, she would ask me where he was, where she could reach him, just as she had done before. She seemed so desperate.

* * *

On March 23, Palm Sunday, Pat Griscom went to the East Coast to visit family. Van drove her to the airport and said that he would be there to pick her up on Easter Sunday when she was to return. However three days later, he told the Fefermans he had to go to Mexico City. He canceled a dental appointment, and changed two dates in Cambridge, Massachusetts, where he had planned to go the following week: one with his tax lawyer and the other with his friend Burton Dreben, who had invited him to a Harvard Society of Fellows dinner on March 31.

Atypically, Van asked for a ride to the airport, and on the way he explained—half to himself, it seemed—why he felt compelled to go: "Adolfo called me. He said if I don't come, Anne-Marie will kill herself. Of course, she wants to kill *me* too." His tone was matter-of-fact, as if he were telling old news. Disbelieving, his friends said, "But if you really think she means to kill you why are you going?" Van answered, "Because Adolfo called me. I must go. And *she* called me too." Unspoken was the assumption that his presence in Mexico would calm Anne-Marie and prevent her from harming herself.

Later, there were reports that Anne-Marie had warned her family she would kill herself on Good Friday, but Van did not mention that in the car. He did however repeat what he had said before to Pat and others: "Anne-Marie has guns; she carries them as a matter of course. She never goes anywhere without a gun. Every time I go down there, I have to take them away from her."

* * *

Carrying his income tax report and a few *Gödel* notes with him, in the hope that he might find an odd moment to work, Van arrived in Mexico City Thursday night. Saturday afternoon or early evening, as he lay sleeping in his study (rather peacefully it seems, judging from the photographs taken by the police), his pants folded and his papers neatly stacked beside him on the bed, his wife shot three bullets into his head with her wooden-handled Colt 38 special, put the revolver into her own mouth and shot one more bullet. Her maid, the only other person in the house, said she heard nothing. No arguing, no shots. She found the bodies after ten in the evening when she came to see why the *señora* had not called for dinner. In her report to the police, she said that earlier in the day, when she had brought the newspapers and breakfast, the *señora* had complained that the oatmeal was cold, and she had to take it back and bring some piping hot cereal. The *señor,* joking and apologizing for the fuss, had said of himself, "Oh, that old man!" and they had all laughed.

There is something illogical yet perfectly characteristic about van Heijenoort's eating his usual healthy breakfast with the woman who was about to kill him—a domesticity that matched the folded trousers and the taxes he was preparing for his accountant, but denied Anne-Marie's repeated assertions that she would end her life and his. On the one hand, he took her threat of suicide very seriously; if he had not, he would not have come to Mexico at that point. Yet, in the face of all the evidence of his wife's emotional instability, her alternating hysteria and depression and her excessive doses of drugs, he held on to his own delusion that he and his wife were those "two reasonable people" who could "sit down and come to an agreement." Perhaps he also thought that if necessary he could physically prevent her from carrying out her threats, but as for the manner in which he was actually killed, it would seem that it simply had not occurred to him that she would shoot him while he slept.

The first accounts in the Mexican press, later followed by the international press, described a "double suicide" in the elegant Lomas de Chapultepec district of Mexico City. The ascetic, ex-Marxist, Professor van Heijenoort would have laughed to find himself identified in news releases as "a wealthy businessman," but he would have been deeply troubled by the more serious inaccuracy. It took several days before corrections were made and the cause of death, based upon the conclusions of the Mexico City police, was reported as "homicide–suicide." Peter Katel, then an American journalist for the *Mexico City News,* wrote the first accurate obituary of Jean van Heijenoort, and the other newspapers followed his lead. He also informed the Mexico City police that the "businessman" was an internationally known professor of logic, a United States citizen, and, formerly, Trotsky's bodyguard and secretary. A few journalists persisted in writing of the "mysterious double suicide" while others, recalling Trotsky's assassination, tried to conjure up a political motive.

*The Zamora-Laurent crypt. An intertwined 'Z-L' is faintly visible on the glass door behind the grill.*

*Panthéon Français (French cemetery) in Mexico City.*

Katel was more than casually aware of many details of Jean van Heijenoort's life because he had known him personally since his own childhood. His parents and the van Heijenoorts had been friends in New York, and Peter and Laure had played together as babies. Although Katel lived in Mexico City, he learned of the tragedy in a telephone call from his mother in New York, who had heard it from Bunny, Van's second wife. The news had traveled first to France and then to the United States, and finally back to Mexico.

The Zamora family gave absolutely no information to the press or anyone else other than the police. Nor did they ever communicate or commiserate with the van Heijenoort children about their father's death. Desolated by their loss, Anne-Marie's parents refused to discuss anything related to the tragedy. Four years later, as this book was being completed, they could still barely speak of the event. However, there were reports from friends that the elder Zamoras held their son-in-law responsible for what they continued to maintain was a double suicide.

Certainly Adolfo Zamora knew of Anne-Marie's anguish and of her stated plans to kill herself; he had, after all, summoned Van to Mexico. Also, since she was making no secret of her feelings, it is likely that Zamora would have heard his daughter attribute the cause of her despair to Van's absence. But Zamora had no way of knowing his son-in-law's state of mind, and by this time Zamora probably no longer cared what he was thinking and feeling. In his view, Jean van Heijenoort, his former ally and dear comrade, the "pure" man he had once held up as the paragon of virtue and selflessness, was no longer the same person; for in his attempt to save himself, he had driven Anne-Marie, Zamora's cherished daughter, beyond the edge.

A small, hasty funeral was held on Easter Sunday, 1986, at the Panthéon Française, the French cemetery of Mexico City. The bodies of Ana Maria Zamora Laurent and Jean van Heijenoort were the first to be placed in the extravagant marble crypt reserved for the Zamora-Laurent family. Only the letters 'Z-L' on the door give any indication of whose bones and spirit reside there.

# Afterword

AT THE HASTY, ALMOST furtive funeral in Mexico City that marked the traumatic end of Jean van Heijenoort's and Ana Maria Zamora's life, no time was given to remembrance or tribute. The Zamora family's shock and grief appear to have been beyond words, and, still unaware of his death, no one of Jean's family or friends was present, except for Esteban Volkov, Trotsky's grandson. But in the following months, at Stanford University, Harvard University, and the Institut Henri Poincaré in Paris, van Heijenoort's colleagues had their own memorial gatherings to honor him and to express their feelings of affection and loss. I had seen Van at such gatherings held in the memory of others, so I know he would have been moved and probably amused to have more than one community claim him as their own.

The three ceremonies were linked by common threads: all were in academic settings; at each one, individuals rose and, as if with a single voice, praised van Heijenoort's outstanding intellectual contributions and his devotion to scholarly pursuits; his unusually high standards of perfection and his concern for the work of others; his modesty and the spare elegance of his language. Yes, everyone agreed, not without a hint of annoyance, this man rarely said one word more than was absolutely necessary. He was private, elusive, and prismatic; no one had seen the whole of him.

There were, of course, differences in tone among the three memorials. In Paris, the largest gathering, there was a strong political strain; Van's revolutionary past was in the foreground. As Jean-Yves Girard put it, so succinctly, "A lot of old Trotskyites were in the hall." Indeed, the number of political comrades exceeded the academic colleagues, an imbalance that troubled some of the logicians. And even though the friends who spoke of Van's years with Trotsky knew perfectly well that he had long since abandoned Marxism (in fact, many were themselves no longer of the Left), their eulogies were so like a paean to a fallen comrade that a visiting American philosopher, ignorant of van Heijenoort's current political leanings, came away convinced that he had died a Trotskyite. Only those with a wide-angle view of Van's life understood these deeply emotional bonds which, like blood-ties, were never completely severed. For better or worse, this was family.

At Harvard, although politics was certainly part of the picture, the emphasis was more academic and, at the same time, more intimate. Van's son and daughter (they attended all the memorials), his second wife, at least one former lover, and many of his closest personal friends and colleagues were present. It was there, after the more formal talks, that Douglas Bryant said, ruefully, what many others were thinking: "How ironic that John should die like that, after all those years with Trotsky, living in real danger, with a price on his head." But the women who called Van "Jean" knew better; neither they nor I were surprised that domestic, indeed romantic, rather than political, passion had killed him. To my mind, one of the saddest ironies of his dying by Anne-Marie's hand was that he had struggled so long and hard to effect a reconciliation because he was convinced he could not live without her, only to realize, again, that it was impossible to live *with* her.

Van had a way of saying with a resigned sigh, "Nothing in my life is simple; everything is complex." It sounded like a complaint, but I think he liked it that way. He took fabulous care of his physical self and yet he flirted with death. What else can explain his going to the woman who had repeatedly said she wanted to kill him? But I can also hear him saying: Suicide? No, that's absurd. I can't die yet. I have too much work to finish.

The exterior van Heijenoort was a logician, an eminently reasonable man who met all his obligations on time. The interior person living under the same pallid skin was impulsive, irrational, and driven. If there was nothing preordained about van Heijenoort's violent death neither was it inconsistent with his life.

Less than a week before he died, Van had dinner at my house, and then we went to see *Hannah and Her Sisters.* Nothing about the evening was unusual: not the modest meal with just the three of us – Sol, Van, and me – not our conversation about food, work, travel plans, and movies – especially Woody Allen movies. Only in retrospect did I take note that our dinner had been on Palm Sunday, that he died on Good Friday and was buried on Easter Sunday. The coincidence is haunting, because in spite of the secularity of his life, there was something deeply religious–some have even said saintly – about him.

Van loved good movies, and, not surprisingly, Woody Allen as auteur and actor was one of his great favorites. Woody playing the indefatigable neurotic hypochondriac in *Hannah and Her Sisters* suited Van to a tee; Woody, repeating "I can't sleep; I'm dizzy; I'm dying!" as he visited a series of doctor-diagnosticians; Woody hooked onto and into multifarious scan machines. Van laughed louder than anyone else in the theatre, a howl of self-recognition. The anxieties of complicated family relationships, the professor-student infatuations, the impossible obssessive love, he knew them all.

None of these problems was mentioned explicitly during our last evening together, nor did we speak of Anne-Marie. However, two days later, in response to Adolfo Zamora's telephone call, Van made the fatal decision to fly to Mexico City "just for a few days," he said. He would be back at Stanford by Easter, and then go to Cambridge, Massachusetts the following week in plenty of time to do his taxes before the April 15 deadline. For all his wandering, Van took great pains to let his friends know where he would be and when. As Anne Fagot-Largeault once said, "You always expected him to turn up, and you were always glad when he did."

Only this time, he did not.

# Appendix

## JEAN VAN HEIJENOORT'S SCHOLARLY WORK
### 1948–1986

by

SOLOMON FEFERMAN

IN 1948, DURING THE SAME year that Jean van Heijenoort published "A Century's Balance Sheet" in *Partisan Review* (under the pseudonym Jean Vannier), he wrote a long piece entitled "Friedrich Engels and Mathematics," which did not appear in print until the publication of his *Selected Essays* (1985).[1] During that same period, van Heijenoort was also working on a dissertation in mathematics at New York University, which he was to complete in 1949. It is interesting to compare the piece in *PR* with that on Engels, as together they mark the end of his adherence to the Marxist cause and signal the beginning of an academic career devoted to critical thought.

"A Century's Balance Sheet," van Heijenoort's first public break with Marxism, appeared on the centennial of the publication of Marx and Engels' *Communist Manifesto.* The main point of his article was to criticize the supposed scientific certitude of Marx's fundamental hypothesis about the inevitable rise of the proletariat, in view of subsequent developments. The tone is polemical and there are few specifics, but the call is for a "rational and methodical scrutiny of the lessons of the past and of present possibilities." In contrast, the second piece is narrowly focused on Engels' mathematical knowledge—a different question on the face of it—but the underlying message is the same. It begins:

> Friedrich Engels has passed judgment on many points in mathematics and its philosophy. What are his opinions worth? Important in itself, this question has a more general interest, for Engels' views on mathematics are part of his "dialectical materialism," and their examination gives a valuable insight into this doctrine.

This is followed by a thoroughly documented and devastating critique, demonstrating Engels' puerile knowledge of mathematics and the pretentiousness of his pronouncements on that subject and their supposed significance for "dialectical science." It concludes with a brief assessment of Marx's own efforts to extract from mathematics something relevant to his theories, efforts which van Heijenoort placed "well above those of Engels" but still considered to be "no better than those of an alert student of the calculus."[2] Taken at face value, the essay may be considered a bit of overkill, but it served to elaborate van Heijenoort's critical reassessment of Marxism and his rejection of it on theoretical scientific grounds. Implicitly, he was saying: if its founders held knowledge and opinions like these, how can we continue to take Marxism seriously?

The Engels piece was placed last in the *Selected Essays* although it was written twenty years before the first of the other contributions. In his Foreword to that volume, van Heijenoort said: "Were I to write on this topic now, I would do it quite differently; but I let this old article of mine stand as it is." Evidently he was not too much embarrassed by this essay for it prefigures the care and thoroughness that he was later to devote to his work in the history and philosophy of logic. Moreover, by way of demonstrating what Engels could (and should) have known if he had been at all aware of the general and well-publicized development in the mathematics of his time, van Heijenoort gives a résumé of such nineteenth century topics as: non-Euclidean geometry, the foundations of the calculus and elimination of infinitesimals, the legitimation of "imaginary" numbers, the emerging mathematics of the infinite, the role of logic in mathematics, and the relationship between the idealizations of mathematics and its application to empirical science. Here indeed was a vigorous warm-up for his later scholarly contributions, already demonstrating considerable independent study and reading.

Van Heijenoort had concluded his school career in 1930 with the unusual award of a double *baccalaureat* in mathematics and philosophy. Logic forms a natural bridge between these disciplines for many good reasons, and there is evidence that he took a special interest in this subject well before returning to his formal studies in 1945 (for example, by beginning to grapple with *Principia Mathematica,* the monumental work by A.N. Whitehead and Bertrand Russell, which attempted to give a logical foundation in purely symbolic terms for all of mathematics). In its relation to such earlier interests and to his subsequent career, van Heijenoort's dissertation work at NYU may appear anomalous until one recalls his rigorous training in classical mathematics as an undergraduate at the Lycée Saint-Louis in Paris. The title of his thesis, written under the direction of Professor J.J. Stoker, and accepted in 1949, was *On Locally Convex Surfaces.* It fell within the field of topology, although it was closely related to earlier results in differential geometry. Its main theorem gives a rather general sufficient condition for a two-dimensional surface (manifold) embedded in Euclidean 3-space to be the boundary of a convex body. This turned out to have been discovered independently in 1948 by different methods, by the Soviet mathematician A.D. Alexandrov.[3] However, working hastily, van Heijenoort realized that his method could be generalized to higher dimensional manifolds in a way that Alexandrov's apparently could not. He published his thesis work in detail for the two-dimensional case with indications of how it could be generalized to higher dimensions, in a 1952 paper, "On Locally Convex Manifolds." For this work, van Heijenoort anglicized his first name to 'John,' as was his custom at the time.

Since the graduate mathematics center at NYU was almost wholly devoted to the subject of classical analysis and its scientific applications, van Heijenoort would not have been able to write a thesis in mathematical logic, even had he been clearly of a mind to do so. Indeed there were few universities in the United States at that time where it was possible to do graduate work in logic, the principal ones being Harvard, Princeton, the University of California at Berkeley, and Wisconsin. At any rate, while his thesis did not have significant impact and he did not continue that line of reasearch, it was a solid and original contibution which was still assessed in positive terms years later by experts in the field.[4]

Van Heijenoort's deepening understanding of logic came through his own studies in the 1950s and 1960s, reinforced by what little teaching he could do in this subject. When he was a faculty member in the undergraduate school of NYU, he taught the basic courses in mathematics, and only every other year did he give a course in logic. He had larger and more receptive audiences in 1962–1964 when he was a visiting professor of philosophy at Columbia University. The following year, with his appointment as professor of philosophy at Brandeis University, he was able to devote himself exclusively to the teaching of logic and associated philosophical topics. He continued to do so until his retirement in 1977, although the development of modern logic as a principal line of study at Brandeis was not expanded as he had hoped.

Through his gradual contact with logicians in the 1950s, and especially via the 1957 Cornell Summer Institute in Symbolic Logic, van Heijenoort was drawn into the circle of reviewers for *The Journal of Symbolic Logic,* whose review section was at that time (as it had been from its inception in 1936) under the editorship of the noted logician Alonzo Church. Church's aim for that section was to give a complete and thorough critical report of the mathematical and philosophical literature in symbolic logic, an endeavor to which he himself contributed substantially, first by amassing a bibliography of such works up to 1936, and subsequently by his own frequent reviews.[5] Besides being known for his original contributions to logic, Church was noted for his great precision with respect to all aspects of the subject, be they mathematical, philosophical or historical. This accorded with van Heijenoort's temperament, and his service as a reviewer for *The Journal of Symbolic Logic* helped sharpen the tools for what were to become his major contributions. In all, he was to write forty-three reviews for the *Journal* in the period 1956–1975, thirty-three of which appeared prior to 1967, the year of the publication of *From Frege to Gödel.* The majority of these were of historical, philosophical, or expository articles and books, but some also concerned technical logical work on matters he found of personal interest. Naturally, as his own work took over, he had less time and inclination to write reviews.

Following his collection of materials in the 1950s for autodidactic and pedagogical pruposes, van Heijenoort began working on a source book in mathematical logic for the Harvard University Press series of *Source Books in the History of Science,* at the behest of Professors Burton Dreben and Willard Van Orman Quine of the Harvard philosophy department. He was almost totally immersed in this project for six or seven years, but the detailed consideration of the literature the work required would provide him with material to be mined for a lifetime. Within a few years of taking on the project, van Heijenoort contributed a pair of articles to the eight-volume *Encyclopedia of Philosophy* (1963), edited by Paul Edwards. The topics of those articles are of great significance for the period 1879–1931 to which van Heijenoort was devoting his source book. One of the articles in question provides a survey of the best known logical paradoxes, including some having their origins in antiquity, but with special attention to those which sprang up in the modern period. The most famous and the simplest of these paradoxes was that discovered by Russell in 1901 which uncovered a fatal defect in the system for the logical foundations of mathematics promulgated by Gottlob Frege, beginning in 1879. Russell then struggled to repair Frege's program, and for this purpose he eventually settled on the "theory of types" which was to form the basis of *Principia Mathematica.* Another approach dealing with Russell's paradox, as well as an earlier paradox found by Georg Cantor in his informal theory of sets, was provided by Ernst Zermelo's axiomatic theory of sets.

The logical paradoxes hung like a shadow over the foundational discussions in the first third of the twentieth century, and partly out of the fear of possible inconsistency at the very basis of mathematics, radical programs emerged. Among the proposed solutions was that advanced by David Hilbert, whose aim was to establish the consistency of mathematics by formalizing it in completely precise symbolic form and then demonstrating by the most rudimentary, "finitary" means of mathematical reasoning that no rigorous chain of deductions could lead to a contradiction. The second of van Heijenoort's articles for the *Encyclopedia of Philosophy* was about Kurt Gödel's famous incompleteness results of 1931, according to which no sufficiently strong consistent formal system is complete, and moreover its consistency cannot be established by means which can be formalized within the system itself. It was this latter result which seemed to undermine Hilbert's program, and constituted a watershed in the development of logic in this century. In his article, van Heijenoort gave a clear and careful outline of Gödel's proof, followed by a full discussion of the influence and significance of his incompleteness theorems.

Van Heijenoort's major scholarly contribution was his work as editor for the volume *From Frege to Gödel: A Source Book in Mathematical Logic, 1879–1931.* Some background is needed to help appreciate his achievement. Logic has always been of interest to philosophers and mathematicians, but until the nineteenth century it was more the province of the former and in fact had been dominated by the Aristotelian classification of syllogistic reasoning. Around the mid-nineteenth century, logic was taken up by the mathematicians as a subject to be developed like other parts of mathematics, under the heading "the algebra of logic." The leaders in this transformation were George Boole, Augustus DeMorgan, W.S. Jevons, Charles S. Peirce, and Ernst Schröder. Boole developed a formal *calculus of classes,* which subsumed Aristotelian logic, whose abstract laws of union, intersection, and complement (written as +, ×, and −, respectively) came to be called Boolean algebra.[6] Peirce and Schröder expanded this algebra to provide a *calculus of relations.* During the same period, the process of clarification and rigorous systematization of fundamental mathematical concepts was reaching deeper, to the foundation of the basic number systems themselves, through the work of Georg Cantor and Richard Dedekind for the real numbers ("measurement numbers") and of Dedekind and Giuseppe Peano for the even more basic system of the positive integers (the counting numbers 1,2,3,. . .). Cantor also developed a controversial extension of the concepts of cardinal and ordinal numbers and of arithmetic operations on them to include "transfinite" numbers, with surprising and striking consequences, but his justification for their use involved problematic assumptions about infinite sets.

A revolutionary new step in logic came from Gottlob Frege with his publication in 1879 of *Begriffsschrift* (literally, "concept writing" or ideography), which introduced a new and very general symbolism for expressing mathematical concepts. Frege's logic of predicates (properties or relations) as combined by the propositional connectives ('and,' 'or,' 'not,'. . .) and quantifiers ('all' and 'some') established the laws of complex interrelated universal and existential statements, now called *quantificational* or *predicate logic,* or the *predicate calculus.* Thinking mathematically, Frege identified predicates with propositional functions which express specific propositions for each choice of arguments. Quantification over arbitrary propositional functions could also be made in his symbolism, as was necessary for Frege's aim of defining all mathematical concepts, beginning with those of the integers, in purely logical terms, and of deriving their basic properties from purely logical principles; that aim is called the *logicist program.* Coincidentally, Peano was working independently with a symbolic language within which to express the basic principles of the positive integers, now known as Peano's Axioms; however, his language was not organized as systematically as that of Frege, nor did he plumb so deeply.

The paradox announced by Bertrand Russell in 1901 showed that Frege's plan (as spelled out in 1879 and more fully in 1893) was seriously defective, since by allowing functions or classes to be dealt with on the same plane as individuals it led to the possibility of contradictory self-reference ("the class of all classes which do not contain themselves as members"). Russell's own effort in the *Principia Mathematica* to avoid the paradoxes by making rigid type distinctions was only partially successful, since he had to make an *ad hoc* assumption (the "Axiom of Reducibility") to carry the program through. As a result, his enormous effort is generally considered to have failed. The logicist program was then followed by other competing programs for the foundations of mathematics, including Hilbert's *finitist program* for establishing the consistency of formal systems for mathematics via a *theory of proofs* and Zermelo's *set-theoretical foundations* (later expanded by Abraham Fraenkel). Another radical approach was promulgated by the mathematician L.E.J. Brouwer under the name of *intuitionism* (a form of constructivism). Brouwer rejected the use of the Aristotelian Law of the Excluded Middle ("A or not-A") applied to statements about "completed" infinite totalities such as the totality of integers or of real numbers, and he initiated a reconstruction of mathematics on entirely constructivist grounds, though at the cost of introducing some new problematic concepts and principles.

The mathematicians in Hilbert's school, in the 1920s, such as Wilhelm Ackermann, Paul Bernays, and John von Neumann contributed to his consistency program, but the deepest advance in the theory of proofs during that period was made by Jacques Herbrand in 1929. In the same year (in his doctoral thesis), Kurt Gödel obtained the fundamental result establishing the completeness of the first-order predicate calculus, a problem that had been posed by Hilbert and Ackermann in their influential logic text of 1928. Most of the tools for this *completeness theorem* were already available in the work of Thoralf Skolem,though he had not faced the completeness problem squarely. But Gödel's *incompleteness theorems* of 1931 (for systems containing Peano's axioms) involved striking new techniques and served not only to put Hilbert's program in question but to usher in a whole new era in mathematical logic, to wit, that of *metamathematics*, the systematic study of formal languages and formal systems for mathematics by mathematical means themselves. The 1930s were dominated by developments in the theory of *effective computability* (also known as the theory of *recursive functions*), with results about effectively unsolvable problems by Alan Turing, Stephen C. Kleene, and others, using proposed definitions of effectiveness that had been advanced by Church, Herbrand, Gödel, and Turing. That period also saw substantial progress made in proof theory by Gerhard Gentzen, in the semantics of formal languages by Alfred Tarski and his subsequent "theory of models," and in axiomatic set theory, again at the hands of Gödel. Still, although the Association for Symbolic Logic was founded in 1936 and its *Journal* drew a respectable and steady number of articles in the years following, the field of logic did not really take off until after World War II.

Logicians of the postwar generation were only dimly aware of the history of their subject, and historical study was not integrated with their current knowledge and endeavors. There was one book, *The Development of Logic,* by W. and M. Kneale that began with the Greeks and worked up to Gödel's results, but it was better on the ancient and pre-modern period than on the early twentieth century. Some historical information on different aspects of the subject could also be gleaned from the texts of Kleene (1952), Church (1956), and Fraenkel and Bar-Hillel (1958). Then, in 1965, Martin Davis published *The Undecidable,* a collection of basic papers, spanning the period 1931–1944, by Gödel, Church, Turing, Kleene, and others on the incompleteness and unsolvability phenomena and recursive function theory. In that volume, some of Gödel's papers were translated from German into English for the first time, and Davis provided very brief introductions to the individual articles. But, except for the volume by Kneale and Kneale, all such historical presentations were written by logicians for logicians and did not involve serious historical work.

With this background in mind, both the nature of van Heijenoort's accomplishments in his *Source Book* and the impact it had can be appreciated. With the assistance of the Harvard logicians Burton Dreben, W.V.O. Quine, and Hao Wang, he selected forty-six of the most important texts (articles, monographs, books, and letters) in mathematical logic and the foundations of mathematics published in the period 1879–1931, for reproduction in whole or part in careful English translations. The book begins with Frege's *Begriffsschrift*, given in full, followed by texts of Peano, Dedekind, Cantor, Russell, Hilbert, Zermelo, Skolem, Fraenkel, Brouwer, von Neumann, Bernays, Ackermann, Herbrand, and Gödel, among others. The final four major items are Chapter Five of Herbrand's 1930 dissertation *Investigations in Proof Theory*, Gödel's 1930 paper "The Completeness of the Axioms of the Functional Calculus of Logic," his 1931 incompleteness paper "On Formally Undecidable Propositions of *Principia Mathematica* and Related Systems," and, at the end, Herbrand's 1931 paper, "On the Consistency of Arithmetic." The originals of the selections for the *Source Book* were in seven different languages (English, Latin, French, German, Italian, Dutch, and Russian). The majority of the translations were made by Stefan Bauer-Mengelberg; a few were done by Beverly Woodward and Dagfinn Føllesdal, and the rest were by van Heijenoort himself. Each text was preceded by a substantial introductory note, giving historical data and background for the item, explaining something of its contents, elucidating any problematic points, and, finally, pointing to later work that was influenced by the work in question.

The service which the *Source Book* provided to working logicians (especially the English-speaking community) in making available and putting so many important, but largely inaccessible texts into historical context, was unquestionable. As Professor Quine later said:

> In his *Source Book* van Heijenoort made a contribution that was second only—as an editorial contribution—to what Church had contributed in molding *The Journal of Symbolic Logic* itself.[7]

But besides its value to the community of logicians, the accomplishment of the *Source Book* from the point of view of scholarship was to raise the study of the history of modern logic to a new level by its evident exercise of the highest historical, logical and expository standards.

This is not to say that *From Frege to Gödel* was acclaimed without reserve, especially in regard to the principles of selection. On this point, van Heijenoort had said in his preface:

> The second half of the nineteenth century saw a rebirth of logic. ... Boole, DeMorgan, and Jevons are regarded as the initiators of modern logic, and rightly so. The first phase, however, suffered from a number of limitations. It tried to copy mathematics too closely, and often artificially. ... Considered by itself, the period would, no doubt, leave its mark upon the history of logic, but it would not count as a great epoch. ... A great epoch in the history of logic did open in 1879, when Gottlob Frege's *Begriffsschrift* was published. This book freed logic from an artificial connection with mathematics but at the same time prepared a deeper interrelation between these two sciences. It presented to the world, in full-fledged form, the propositional calculus and quantification theory. ... The next decades saw striking advances in logic. Two new fields, set theory and foundations of mathematics, emerged on the borders of logic, mathematics, and philosophy.

As to the choices to be made "so as to depict the new development," van Heijenoort said that "various conditions, sometimes hard to reconcile" had to be satisfied. "The main constraint was that the texts to be selected had to fit between the covers of a single volume. This precluded encyclopedic completeness; the book had to hold to the main line of the development." Despite this rationale, *From Frege to Gödel* was criticized by Gregory H. Moore in a review (in 1977) principally on the criteria for selection. He argued that van Heijenoort underestimated the work of Boole and his followers and overestimated that of Frege, so that fully one-eighth of the *Source Book* is given over to Frege's *Begriffsschrift.* Moore further criticized the omission of the critical writings of Henri Poincaré, of excerpts from Brouwer's earlier work, and—"most surprising and least justifiable"—of any contributions by the fine Polish school of logicians that emerged after World War I, especially Stefan Leśniewski, Jan Łukasiewicz, and Alfred Tarski. He did, however, compliment the *Source Book* "for its many translations and excellent introductions."

A much more enthusiastic response had been expressed by the distinguished logician Paul Bernays in his review of *From Frege to Gödel* in the *Journal of Philosophy* (1970). He found the comments given in the introductory notes and in additional footnotes especially valuable, providing as they do "... helpful explanations, discussions, and plenty of historical information, whose concern goes far beyond the limit of 1931, nearly to the present." He pointed out that almost all authors with works of relevance to this period are included in a comprehensive bibliography at the end of the volume. (In fact, there one will find listings stretching from the 1600s to the mid-1960s.) Bernays also observed that the introductory notes serve to provide information about papers not included in the volume; for example, the theory of relations of Peirce and Schröder was outlined in the introduction to the article "On Possibilities in the Calculus of Relatives" by Leopold Löwenheim. In Bernays' view:

> On the whole the book ... furnishes characteristic representatives of the epoch of logic and foundations of mathematics mentioned in its title, and thereby recalls in an authentic way most of the mental steps accomplished in this epoch: the discovery of the set-theoretic antinomies, the precising [sic] of the axiomatic method, the constitution of axiomatic set theory, the enterprises of the *Principia Mathematica,* of Brouwer's intuitionism, and of Hilbert's proof theory, and the discovery of the restrictions of formalized axiomatics by the results of Skolem and Gödel. It likewise affords an impressive survey of the development of foundational investigations and of their rich results.[8]

Two of the most famous texts from the *Source Book*—its poles, so to speak—were Frege's *Begriffsschrift* and Gödel's 1931 incompleteness paper. These, in their English translation together with the introductory notes (and some minor corrections) were extracted from the larger volume for publication as a separate edition in 1970, *Frege and Gödel: Two Fundamental Texts in Modern Logic,* again published by Harvard Press. The original volume went into four editions and was eventually printed in paper back.

With the appearance of the *Source Book,* van Heijenoort became well-known as an historian of logic: he was made an editor of *The Journal of Symbolic Logic* in 1968 and served on its editorial board until 1979. He also became an editor of the *Archives for the History of the Exact Sciences,* and later a member of the editorial board of the series in the *History of Logic* published by Bibliopolis in Naples, Italy, serving in both capacities until his death in 1986.

It was also in 1967 that the first of a series of essays were published (the remaining ones were from 1973 on) in which van Heijenoort gave more wide-ranging reflection to the major transformations in logic from the mid-1800s on. The 1967 essay, "Logic as Calculus and Logic as Language," has been referred to as a classic.[9] The point of departure is the controversy between Schröder and Frege about the relative merits of the algebra of logic and Frege's quantification theory, and in particular as to which fulfilled Leibniz's dreams of a *calculus ratiocinator* (in which all disputes would be settled by routine calculation) and of a *lingua characterica* (a universal symbolic language). Expanding on this point from his introduction to the *Begriffsschrift* in the *Source Book*, van Heijenoort clearly sides with Frege, putting him on the "universal language" side and Schröder and company on the "calculus" side of logic (insofar as that went). In his introduction he had lauded Frege's monograph as "perhaps the most important single work ever written in logic," especially for its formulation of quantification theory. In the present essay he stressed, in addition, the importance of Frege in establishing the modern notion of formal system and, as a philosopher, in inaugurating the analytic tradition. The main part of the essay, though, concerns the historical shift from Frege's conception of his logic as an all-encompassing system, to the modern view, inaugurated in model theory by the Löwenheim-Skolem theorem and in proof theory by the work on Hilbert's program, under which logical systems are treated locally and from the outside (thus "metamathematics"). This is a theme to which van Heijenoort would return frequently. Moreover, despite his enormous admiration for Frege's accomplishments and the obvious attractions of universal absolutistic systems of thought, van Heijenoort would steadily ally himself with a more relativistic and tentative point of view; this resonates with his passage through Marxism, with its supposed answers to all economic and societal questions, to a more skeptical point of view about universal solutions of any kind.

Within a year of the appearance of the *Source Book*, van Heijenoort produced a complete edition of the logical writings of Jacques Herbrand, *Écrits Logiques* (1968), to which he contributed a long preface. There he lamented the many tragic losses to logic in France in the first half of the twentieth-century, saying, "*Un mauvais sort semble s'être acharné sur la logique en France.*"[10] In 1914, Couturat died in an automobile accident; in 1924, Nicod, at thirty-one, died of an illness; in 1931, Herbrand, only twenty-three, died in a mountain-climbing accident; and in 1944, Cavaillès and Lautmann were killed by the Germans for taking part in the resistance movement. (Logic was not actively pursued as a subject in France until the 1950s, but since then it has thrived.)

In his short life, Herbrand produced important basic results both in logic and in pure number theory, for which van Heijenoort calls him a "*génie créateur.*" Herbrand's fundamental theorem from his 1930 dissertation gives a kind of reduction, in the limit, of quantificational logic to propositional logic. This has led to considerable further work in proof theory, among which the most noteworthy is that of Gerhard Gentzen (beginning in 1934); it has also been used in recent years as the basis for one approach to automated deduction. However Herbrand's own proofs were flawed; already in 1939 Bernays remarked that "Herbrand's proof is hard to follow," and in 1943, Gödel uncovered an essential gap, though his notes on this were never published. Counter-examples to two important lemmas in Herbrand were produced by Burton Dreben in collaboration with his students and colleagues Peter Andrews and Stål Aanderaa in an article in 1963, in which they outlined how the arguments could be repaired. A detailed proof of the crucial lemma was later given by Dreben and John Denton in 1966.[11]

Despite the known flaws, van Heijenoort regarded Herbrand's theorem as one of the deepest results of logic. (In the minds of most logicians that assessment is debatable, but the importance of the theorem is undeniable.) Because of this view – and perhaps because of the French connection, Herbrand was of permanent interest to van Heijenoort. He wrote an extended essay (in French) in 1981, "L'oeuvre Logique de Jacques Herbrand et son Context Historique" and a revised English version in 1985 ("Jacques Herbrand's Work in Logic and its Historical Context"), that appeared in his *Selected Essays.* In the last years of his life he was also engaged in a project with Burton Dreben and Claude Imbert (professor of philosophy at the École Normale Supérieure) to produce an edition of the complete works of Herbrand under the sponsorship of the French Ministry of Culture. Since van Heijenoort's death, for lack of his driving force, the project has gone into limbo.

In the same year as the 1968 edition of Herbrand, van Heijenoort prepared a typescript, "On the Relation Between the Falsifiability Tree Method and the Herbrand Method in Quantification Theory," and in 1975 he went further into a technical account of the fundamental theorem in a typescript simply entitled "Herbrand." Between 1972 and 1975 he prepared a series of expository notes on the related falsifiability tree method in quantification theory for distribution to his students and participants in his courses and seminars at Brandeis University. Irving H. Anellis has reported on these in brief in "Some Unpublished Papers of Jean van Heijenoort" and at length in his book on van Heijenoort's scholarly work.

All of van Heijenoort's technical expositions, both published and unpublished, are characterized by their precision, clarity, and economy of language. The monograph *El Desarrollo de la Teoría de la Cuantificación* (1976) is a compact summary of lectures that van Heijenoort gave at the Universidad Nacional de México in 1973, on four different approaches to quantification theory and their historical development: the axiomatic approach from Frege and Russell to Hilbert and Ackermann, Herbrand's approach, Gentzen's sequential systems and, finally, the natural deduction systems of Jaskowski and Gentzen. His concerns in these lectures are more methodological than historical (cf. the review by José Ferrater Mora (1980)). Later the falsifiability tree approach to completeness of quantification theory, a favorite of van Heijenoort's, was expounded at length for non-classical logics (intuitionistic and modal) in his 1977 lectures at l'École Normale Supérieure de Jeunes Filles, and published as the 1979 monograph *Introduction à la Sémantique des Logiques Non-classiques.* The essays that Jean van Heijenoort wrote between 1973 and 1985 — all but one of which appear in his *Selected Essays* (1985) — concern, in his words, "the place of logic in the body of human knowledge." Roughly speaking, they fall into two categories: those having to do with the history and philosophy of modern logic, per se, and those concerned with the wider significance of logic for philosophy. In the former category one might count "Logic as Calculus and Logic as Language" (1967a), "Subject and Predicate in Western Logic" (1974), "Set-theoretic Semantics" (1976a), "Absolutism and Relativism in Logic" (1979b), the essay on Herbrand's work discussed above (1982), and finally, the posthumous article "Système et Métasystème Chez Russell" (1987). In the latter category one would include "Sense in Frege" (1977), "Frege on Sense

Identity" (1977a), "Ostension and Vagueness" (1979a), and "Frege and Vagueness" (1985a). A recurrent theme in the former group of essays is that of *logica utens* vs. *logica magna*, i.e. logic as a tool for use in various relative or specific situations or structures (Schröder, Löwenheim, Hilbert, etc.) vs. logic as a universal all-encompassing system (Frege, Russell, etc.) As discussed above, van Heijenoort clearly allies himself with the former direction of work, whose successes in mathematics are, in any case, more clear-cut. Concerning some of the efforts to use logic outside of mathematics, for example as applied to the syntax and semantics of natural languages, in his essay "Set-theoretic Semantics" (1976a), van Heijenoort points more to the many difficulties (e.g., *re* the treatment of mass nouns or vague predicates) than to the successes, while nevertheless expressing a hope for the latter. As he writes in his essay:

> The problem is to pass from the local treatment of the semantics of a natural language to a global one. One feels that it is possible and that it should be done. ... Unless God-given, syntactic rules must be grounded somewhere. And where, if not in meaning? But between the exact working of a natural language, with the convolutions and intricacies of its meanings and the logical skeleton representable in quantification theory there is a large gap, in which we are still groping.[12]

Of the four essays—all relatively brief—having a more philosophical character, three deal with Frege's theory of meaning (those dated 1977, 1977a, and 1985a), while the fourth (1979a) is indirectly connected with that subject. In 1975, at Brandeis, van Heijenoort had offered a seminar on Frege's logical and philosophical works, and of course he had been preoccupied with Frege in one way or another for many years. Moreover, all but the first of the four doctoral students he had at Brandeis (Dale Gottleib, Steven F. Savitt, George Graham and Irving H. Anellis), wrote dissertations in mainstream philosophy.[13] It is thus not surprising that van Heijenoort should have ventured into this territory in his own writing.

While van Heijenoort's more purely philosophical essays have not had nearly as much impact as his work in the history and philosophy of logic, they should not be neglected.[14] As Feferman and Feferman wrote in *Logic Colloquium '85*:

> Typically these papers are directed to a few significant points, but contain along the way many *aperçus*. The style is unhurried but succinct, precise but unfussy, clear and graceful; the point of view is definite, but the perspective is balanced. The reader who discovers these papers in the *Selected Essays* will become aware of a side of van Heijenoort ... that has not been sufficiently nor widely enough appreciated. These essays make abundantly evident his many excellent qualities which, in their combination, made him unique.[15]

The scholarly work which engaged much of van Heijenoort's energies in the last four years of his life was as an editor on the project to publish a comprehensive edition of the works of Kurt Gödel, considered by many to be the most important logician of the twentieth century. Besides a great deal of editorial work on the first two of the four scheduled volumes of this edition, van Heijenoort assisted with the translations and contributed, in collaboration with Dreben, an important introductory note to Gödel's 1929 thesis and his completeness paper of 1930. Volume I appeared in 1985, before his death; unfortunately he did not live to see the appearance of Volume II although the work for it was largely completed by 1986. In studying the material uncovered by John W. Dawson from Gödel's *Nachlass*, slated to appear in a later volume, van Heijenoort was greatly pleased to see Gödel's 1943 notes on the gap in Herbrand's proof, but he was most excited by Dawson's discovery of correspondence between Herbrand and Gödel, long believed lost, concerning the fundamental notion of effectively computable function. (The resulting proposal reported by Gödel in his 1934 lectures at Princeton afterwards came to be known as the Herbrand-Gödel definition of general recursive function.)

The prime importance of van Heijenoort's participation as a member of the editorial board for this project was best expressed in the memorial notice which appeared in Volume II of the Gödel *Works*:

> Jean van Heijenoort, our dear friend and co-editor, died on March 29, 1986. His contributions to our work on Kurt Gödel were invaluable in every respect. At the outset, his enthusiastic support was instrumental in our decision to embark upon this project. Then, drawing upon his own extensive editorial experience, he helped us to develop our overall plans as well as to make the

many detailed choices, and throughout the course of the work he devoted himself unstintingly and with the utmost care to whatever task was at hand.

The present volume was largely completed by the time of van Heijenoort's death; indeed he had already begun a detailed examination of some of Gödel's unpublished articles for the succeeding volume. His spirit will continue to animate all our work, and we have taken his standards as our own.

**Acknowledgments:** I am indebted to Irving H. Anellis, Stefan Bauer-Mengelberg, John W. Dawson, Jr., Philippe de Rouilhan, Anita Burdman Feferman, Claude Imbert, David Malament, and Rohit Parikh for their many useful comments on a draft of this appendix.

## NOTES TO THE APPENDIX

1. Jean van Heijenoort. "Friedrich Engels and Mathematics," *Selected Essays.* Naples: Bibliopolis, 1985, pp. 123–151.
2. This corroborates the assessment of an unnamed "distinguished" American mathematician in Appendix B of Edmund Wilson's *To The Finland Station,* New York: Harcourt Brace, 1940, pp. 484–486: "The three articles on mathematics by Marx appear to be the notes of an intelligent (but somewhat befuddled) beginner, who is just starting to understand the meaning of differentiation. I doubt whether they were ever intended as a serious contribution to mathematics. ... Engels has a section on mathematics in his famous *Anti–Dühring* (which is supposed to echo the voice of the Master himself). Unfortunately the mathematics itself is all cock-eyed. It is a pity to build an elaborate philosophical system on a series of gross mathematical errors."
3. According to Irving H. Anellis in his book, *Van Heijenoort: Logic and Its History in the Work and Writings of Jean van Heijenoort.* Ames, IA: Modern Logic Publishing, 1992.
4. For example, by Louis Nirenberg and Richard Schoen.
5. Incidentally, Church's program in the *JSL* for a comprehensive report of all work in symbolic logic had to be abandoned in the early 1980s in favor of a more selective approach, due to the exponential proliferation of the literature beginning in the 1960s.

6. Boolean algebra is still in use today as the foundation of electronic switching circuits.

7. W.V.O. Quine: Memorial meeting at Harvard University on April 28, 1986.

8. Andrzej Mostowski, the Polish logician, wrote another long and very favorable review of *From Frege to Gödel* in the journal *Synthese,* 1968. He too raised some questions about selection. Curiously, the volume was not reviewed as a whole in *The Journal of Symbolic Logic,* but only item by item.

9. By Hans Sluga in his article "Frege against the Booleans," 1987.

10. *"Un Mauvais sort ...":* A terrible curse seemed to have fallen on logic in France.

11. See the introductory note to "Herbrand 1930" in *From Frege to Gödel* for more details. Warren Goldfarb produced an English-language edition of Herbrand's *Logical Writings* in 1971, to which he contributed a long and informative introduction.

12. Jean van Heijenoort, *Selected Essays.* Naples: Bibliopolis, 1985, p. 53.

13. The information about van Heijenoort's teaching and students at Brandeis comes from Anellis' book, *Van Heijenoort: Logic and Its History in the Work and Writings of Jean van Heijenoort.* Ames, IA: Modern Logic Publishing, 1992.

14. It is fair to say that the philosophical essays have had a mixed reception. For two assessments, see the reviews of the *Selected Essays* by David Bell in the journal *History and Philosophy of Logic,* 1986, and by Philippe de Rouilhan in *L'Age de la Science,* 1991.

15. Anita Burdman Feferman and Solomon Feferman, *Logic Colloquium '85.* Amsterdam: North-Holland, 1987, p. 6.

## REFERENCES TO THE APPENDIX


Anellis, Irving H. "Some Unpublished Papers of Jean van Heijenoort". *Historia Mathematica, Vol. 15,* 1988, 270–274.

Anellis, Irving H. *Van Heijenoort: Logic and Its History in the Work and Writings of Jean van Heijenoort.* Ames, IA: Modern Logic Publishing, 1992.

Bell, David. Review of Jean van Heijenoort, *Selected Essays,* in *History and Philosophy of Logic, Vol. 7,* 1986, 226–227.

Bernays, Paul. Review of Jean van Heijenoort (ed.), *From Frege to Gödel,* in *Journal of Philosophy, Vol. 67,* 1970, 109–110.

Church, Alonzo. *Introduction to Mathematical Logic, Vol. I.* Princeton, NJ: Princeton University Press, 1956.

Davis, Martin (ed.). *The Undecidable: Basic Papers on Undecidable Propositions, Unsolvable Problems, and Computable Functions.* Hewlett, NJ: Raven Press, 1965.

de Rouilhan, Philippe. "J.v.H." (En Mémoire de Jean van Heijenoort), in *Logic Colloquium '85.* Amsterdam: North-Holland, 1987, 13–16.

de Rouilhan, Philippe. "De l'Universalité de la Logique." (Compte rendu de Jean van Heijenoort, *Selected Essays*), in *L'Age de la Science, Vol. 4,* 1991.

Dreben, Burton and Jean van Heijenoort. "Introductory Note to *1929, 1930,* and *1930a*", in Gödel 1986, 44–59.

Feferman, Anita Burdman and Solomon Feferman. "Jean van Heijenoort," in *Logic Colloquium '85.* Amsterdam: North-Holland, 1987, 1–7.

Fraenkel, Abraham and Yehoshua Bar-Hillel. *Foundations of Set Theory.* Amsterdam: North-Holland, 1958.

Gentzen, Gerhard. "Untersuchungen über das Logische Schliessen." *Mathematische Zeitschrift, Vol. 39,* 1935, 176–210, 405–431.

Girard, Jean-Yves. "La Mouche Dans La Bouteille." (En Mémoire de Jean van Heijenoort), in *Logic Colloquium '85.* Amsterdam: North-Holland, 1987, 9–12.

Gödel, Kurt. *Collected Works, Vol. I. Publications 1929–1936,* edited by S. Feferman, J.W. Dawson, Jr., S.C. Kleene, G.H. Moore, R.M. Solovay, and J. van Heijenoort. New York: Oxford University Press, 1986.

Gödel, Kurt. *Collected Works, Vol. II. Publications 1938–1974,* edited by S. Feferman (*et al.*), New York: Oxford University Press, 1990.

Herbrand, Jacques. *Écrits Logiques,* edited by Jean van Heijenoort. Paris: Presses Universitaires de France, 1968.

Herbrand, Jacques. *Logical Writings,* edited by Warren Goldfarb. (English translation of Herbrand 1968). Cambridge, MA: Harvard University Press, 1971.

Hilbert, David and Wilhelm Ackermann. *Grundzüge der Theoretischen Logik.* Berlin: Springer-Verlag, 1928.

Kleene, Stephen C. *Introduction to Metamathematics.* Amsterdam: North Holland, 1952.

Kneale, William and Martha Kneale. *The Development of Logic.* Oxford: Oxford University Press, 1962.

Moore, Gregory H. Review of Jean van Heijenoort (ed.), *From Frege to Gödel. Historia Mathematica, Vol. 4,* 1977, 468–471.

Mora, José Ferrater. Review of Jean van Heijenoort, *El Desarrollo de la Teoría de la Cuantificatión,* in *Journal of Symbolic Logic, Vol. 45,* 1980, 635–636.

Mostowski, Andrzej. Review of Jean van Heijenoort (ed.), *From Frege to Gödel,* in *Synthese, Vol. 18,* 1968, 302–305.

Sluga, Hans. "Frege Against the Booleans." *Notre Dame Journal of Formal Logic, Vol. 28,* 1987, 80–98.

van Heijenoort, Jean. (For references to the work of van Heijenoort, see the Bibliography below, pp. 407–408.)

Whitehead, Alfred North and Bertrand Russell. *Principia Mathematica, Vols. 1–3.* Cambridge, UK: Cambridge University Press, 1910–1913.

# Sources and Notes

THE MOST IMPORTANT source for this biography was Jean van Heijenoort himself. In general, Van was a reticent person, yet for this work he was a willing and even talkative subject. Our three years of interviews and conversations provide much of the background and foreground for all but the last two chapters. I trust the reader to keep this in mind and I shall refrain from citing him over and over again. In the instances where I do, I use the abbreviation JvH, and for his son, Jean van Heijenoort, I use JvH *fils*. My name is abbreviated ABF.

**Chapter One: In My Beginning Is My End**

4 "bleed to death": JvH to ABF, 1983.

6 "like a red thread": German expression cited by Stefan Bauer-Mengelberg. Red thread was an identifying mark woven into the ropes of the British Royal Navy, so that their rope could always be recognized.

6 Balagny family history: JvH, JvH *fils*, Paulette Corpion.

9 Dirk van Heijenoort family: Interview March 1990, Delft, the Netherlands.

10 Hélène met Jean Théodore: Paulette Corpion, interviews and correspondence with ABF, 1983, 1989, 1991, Nogent-sur-Oise, France.

14 "casualty of war": Although the family's bitterness about having been abandoned by the medical community is understandable, it is by no means clear that, in 1914, the presence of a doctor would have helped to save Jean Théodore's life. At the time the medical technique for arresting internal hemorrhage was either limited or non-existent. Although the family's attitude toward the event might have been very different, in all likelihood, even with the best of care, Jean Théodore would not have lived.

15 "Dying is not everything": Sartre, Jean Paul. *Les Mots.* Paris, Gallimard, 1965, p. 11. *"Ce n'est pas tout de mourir: il faut mourir à temps. Plus tard, je me fusse senti coupable: un orphelin conscient se donne tort ... Moi, j'étais ravi: ma triste condition imposait le respect, fondait mon importance; je comptais mon deuil au nombre de mes vertus".*

21 "She was perfect": Laure van Heijenoort, interview with ABF, 1983.

## Chapter Two: "Those Schools"

33 "No question of doing his studies": Paulette Corpion, letter, August 1988.

33 "worked like slaves": JvH *fils,* interview with ABF, 1989.

33 "She had never ceased to regret": Interview with Loretta Guyer, New York, 1983.

37 "By one of life's amazing coincidences": Interview with Dr. Françoise Ville, Paris, 1990.

44 "Going to Paris": JvH had been to Paris only twice before. His first visit had been with the parish priest during his brief period of religious intensity. As a reward, the priest had taken the best catechism students to visit Sacré-Coeur, and then, van Heijenoort recalled, "He wanted to take us to lunch, but he was a poor priest from a poor parish, and he didn't have enough money for all of us, so instead, he bought a big bag of cherries and we sat on the steps of Sacré-Coeur and ate them."

## Chapter Three: A Sharp Left Turn

46 "revolution on its wings": JvH to Pierre Broué, 1984.

47 The level of the Lycée Saint-Louis may be confusing to those not familiar with the French system. Although *lycée* is translated as "secondary school," Saint-Louis was the equivalent of a university. Students there were preparing for entrance examinations to the *grandes écoles,* which were essentially graduate schools. Note that the former name of the Lycée Saint-Louis was Collège d'Harcourt, although (as is obvious) the meaning of *collège* is not always the same. For example the Collège de Clermont, for all its rigor, was closer to a secondary school or American high school with a strongly academic emphasis.

50 Bertram D. Wolfe. *Three Who Made a Revolution.* New York: Stein and Day, 1984, p. 189.

51 Paulette Corpion, interview with ABF, June 1983. JvH, interviews with ABF, 1983.

52 At that point, Aragon and Breton had not yet had their rancorous political break, with Aragon following Stalin and Breton moving into Trotsky's camp.

56 Irving Howe. *Leon Trotsky.* New York: Viking Press, 1978, p. 136

56 Jean van Heijenoort. *With Trotsky in Exile.* Cambridge: Harvard University Press, 1978, p. 3.

59 Letter from JvH to Jean-Yves Girard, June 26, 1982:
*"Quant à mon choix entre T. et S., ce fut, je dirais presque, une question de style, de style littéraire même. Il suffisait de lire cinq phrases de l'un et cinq phrases de l'autre. La conviction intellectuelle aussi. J'étais un trotskiste convaincu et militant avant d'aller à Prinkipo, avant de penser que j'irais. Ce n'est donc pas le contact personnel qui a joué au début. Sur un plan moins rationnel, le*

*fait que S. était au pouvoir et T. en exil a certainement été important, car cela correspond à des choses profondes en moi (disons: sensibilité au malheur). Je ne sais pas comment j'aurai réagi en face d'un Trotsky au pouvoir. Mais, d'un autre coté, à Prinkipo, T. ne m'apparaissait pas tellement comme un "vaincu". Jusqu'à l'arrivé de Hitler au pouvoir, son retour [Trotsky's] au Kremlin n'était pas inpensable (penser aux dessous l'affaire Kirov!). Non, le vaincu, c'était S., qui avait abandonné le Leninisme, et l'extermination de tous les vieux bolsheviks durant les epurations fut ressentie par moi comme une défaite de S.*

*"Il y a eu aussi, bien entendu, le hasard. Tres jeune, j'avais rencontré des Trotskistes. Les choses auraient pu tourner différement. Ou l'eussent-elles pu vraiment...?"*

(English translation in text, ABF)

## Chapter Four: The Old Man on the Island of Princes

64 "a towering example": Irving Howe. *Trotsky.* New York: Viking Press, 1978 p. 130.

67 Karl Meyer (pseud. Jean van Heijenoort). "Lev Davidovich." *The Fourth International.* New York: Socialist Workers Party, August 1941, p. 207. (Article written a year after Trotsky's murder.)

67 "gaze": Jean van Heijenoort. *With Trotsky in Exile.* Cambridge: Harvard University Press, 1978 p. 5.

67 "He looks like Otto": Jean van Heijenoort. *With Trotsky in Exile.* Cambridge: Harvard University Press, 1978 p. 3.

70 Trip back to Paris: In November, just a month after Van had arrived in Turkey, Trotsky had gone to Copenhagen to deliver an invited lecture to a Danish student organization and to confer with his political sympathizers. There was even the vain hope that he might find a way to remain in Denmark or somewhere else in Europe. It was during this period that Van brought Sieva to Paris and spent a week there before returning to Turkey with Trotsky.

73 "fine paper": Leon Trotsky. *My Life.* New York: Pathfinder Press, 1970. p. 132.

74 "escape": When the idea of Trotsky's escape was proposed, he raised the objection that his escape would place a double burden upon Alexandra Lvovna. She said, (according to his report) "You must.' Duty to the revolution overshadowed everything else for her, personal considerations especially."

78 "rhythmical and melodious": Jean van Heijenoort. *With Trotsky in Exile.* Cambridge: Harvard University Press, 1978, p. 14.

83 "Not until the exiles were in Mexico": After seven years of total devotion and personal sacrifice, van Heijenoort ended by feeling that if he differed with Trotsky in any degree, Trotsky would cut him off and out of his life without a qualm and with very little emotion. Even if this assessment was wrong, the fact that he thought he was right is painful. But all that poignant reasoning came much later. Natalia Sedova, who one might

have expected to have some special motherly inclination towards young vH, some Russian expansiveness, also seems to have been "all business". In his view, she thought of him as just one of the secretaries, and not as someone special. She considered it a great privilege for anyone to have the opportunity to work with the great man. The idea that someone might have foresworn something else of great merit was irrelevant to her. She was devoting *her* life; why shouldn't others?

84 "I never had enough sleep": Sleep never stopped being an issue with van Heijenoort. He had to have it absolutely quiet. He had to have a gas refrigerator because electric ones were too noisy. He had to be in a quiet neighborhood. He had to be feeling well and not under emotional stress. He and his friend Georg Kreisel frequently compared notes and exchanged addresses about the ideal place to live and kept up a running competition as to who could be the fussiest. First prize usually went to Kreisel.

## Chapter Five: On the Move

88 "think of the thirties as a block.": Among the things which disturbed Jean van Heijenoort most when he spoke of his past and his "political career" was that his "logical" colleagues found it hard to believe that he had been a serious, full time revolutionary and harder still to believe that Trotsky had had even the remotest chance of a return to power.

89 Jean van Heijenoort. *With Trotsky in Exile.* Cambridge: Harvard University Press, 1978, p. 45.

91 Ibid. p. 51.

93 Malraux's meeting with Trotsky: Jean van Heijenoort. *With Trotsky in Exile.* Cambridge: Harvard University Press, 1978, p. 158.

93 Ibid. p. 57.

93 "Exactly": Ibid. p. 57.

95 "the French Collection Institute": Trotsky knew about the Molinier brothers' unsavory business deals from the reports of visitors to Saint-Palais. Van Heijenoort found out about them more directly during a job interview a few years later. When he gave Molinier as a reference, his prospective employer shouted, "What? Who? Get out!" *With Trotsky in Exile.* Cambridge: Harvard University Press, 1978, p. 56.

98 Stavisky scandal: Charlier, Jean-Michel. *Stavisky.* Paris: Robert Laffont, 1974.

100 Jean van Heijenoort. *With Trotsky in Exile.* Cambridge: Harvard University Press, 1978, pp. 64–68.

107 "no creature more disgusting": Leon Trotsky. *Trotsky's Diary in Exile: 1935.* Cambridge: Harvard University Press, 1958.

## Chapter Six: Herring: Norwegian and Red

111 Jean van Heijenoort. *With Trotsky in Exile.* Cambridge: Harvard Univerity Press, 1978, p. 79.

113 Ibid, p. 23–24.

114 André Thirion. *Révolutionnaires sans Révolution.* Paris: Robert Laffont, 1972, p. 409. *"Un jeune Nordique, très grand, un Aryan parfait selon les canons nazis, frais émoulu de la classe de taupe vint aussi demander du travail. Il se nommait van Heijenoort, c'était un des secretaires de Léon Trotski. Il était entré au service du grand homme comme on entre en réligion. Modeste, doux, mais intransigeant et obstiné, intelligent, travailleur, vivant de peu, c'était une sorte d'ange esclave d'une révélation qui le preservait de l'impureté, des hésitations, des doutes et le remplissait de joie ineffable. Il apporta avec lui deux ou trois Trotskyists, Craipeau et une affreuse mégère de vingt ans, ... Elle était amoureuse sans espoir de van Heijenoort."* (English translation in text: ABF)

114 "according to the Nazi canon" seems a perverse choice for describing van Heijenoort but perhaps Thirion was being ironic.

114 The harpie (*mégère*): Pierre Broué claimed that Van knew perfectly well who she was.

115 "*Personellement, j'ai abandonné La Commune ... Toute la tendance de La Commune (qui fut la mienne) est etrangère au marxisme.*" Letter from JvH to Trotsky 16 January, 1936. Trotsky Archive, Houghton Library, Harvard University.

115 "*Quand le fils prodigue revient, on fait cuire un agneau sur le foyer paternel*": Letter from Trotsky to JvH 20 January, 1936. Ibid.

118 André Thirion. *Révolutionnaires sans Révolution.* Paris: Robert Laffont, 1972 p. 374.

119 Gunvor Wraamaan: Interview with ABF, Sept. 1989.

122 Jean van Heijenoort. *With Trotsky in Exile.* Cambridge: Harvard University Press, 1978, p. 91.

122 Irving Howe. *Leon Trotsky.* New York: Viking Press, pp. 130–131.

123 Hayden Herrera. *Frida.* New York: Harper & Row, 1983, pp. 201–204.

## Chapter Seven: A Blue House on a New Planet

131 "would not disembark": Hayden Herrera. *Frida.* New York: Harper and Row, 1983, p. 212.

135 Jean van Heijenoort. *With Trotsky in Exile.* Cambridge: Harvard University Press, 1978, pp. 108–110.

137 Sidney Hook: Interview with ABF, Nov. 1988.

137 Dewey Commission hearings: Jean van Heijenoort, *With Trotsky in Exile.* Cambridge: Harvard University Press, 1978, pp. 108–110. Interviews with JvH, 1984.

139 The hearings of the Dewey Commission were held April 10th to 17th, 1937. The verdict was announced Sept. 21, 1937.

139 "most interesting experience": Letter from John Dewey to Max Eastman, 12 May 1937, Dewey Papers, Lilly Library, Indiana University.

140 "that's exactly my opinion": JvH interview with ABF, 1984.

143 Jean van Heijenoort. *With Trotsky in Exile.* Cambridge: Harvard University Press, 1978. pp. 110–112.

144 Ibid. p. 112.

146 Hayden Herrera. *Frida.* New York: Harper and Row, 1983, pp. 212.

## Chapter Eight: Arrivals and Departures

150 Jvh *fils.* Interview with ABF, 1983.

153 "I should not have said that.": Jean van Heijenoort. *With Trotsky in Exile.* Cambridge: Harvard University Press, 1978. p. 117.

154 Liova's death: Ibid. p. 92, pp. 120–21. Zborowski died in San Francisco in 1990: *San Francisco Chronicle,* May 2, 1990.

158 "Soon afterward": Jean van Heijenoort. *With Trotsky in Exile,* Cambridge: Harvard University Press, p. 125

160 André Breton's speech in praise of Comrade Van, November 11, 1938. (Given to the P.C.I. the Trotskyite group in Paris; reprinted in *Cahiers Léon Trotsky, no. 12,* December 1982.)

*"Le lendemain même de mon arrivée, la joie m'attendait de retrouver là-bas le camarade Van que beaucoup d'entre vous connaissent. Tous ceux qui l'ont approché savent les extraordinaires ressources d'intelligence et de sensibilité qui sont les siennes, ont su apprécier la rapidité de son coup d'oeil, la lucidité de son jugement, mais sans doute n'ont-t-ils pas tous eu le loisir de mesurer l'étendue de sa curiosité ni de faire la part chez lui des admirables dons du coeur. Sa modestie s'offusquerait à coup sûr de mes paroles et cependant je m'en voudrais de laisser passer l'occasion de lui adresser mon salut vraiment fraternel. Qu'il me pardonne de révéler ici ce que son existence présente de pathétique: à tant d'intellectuels qui cherchent dans la négation, dans le saccage de toute conscience morale le secret d'une vie confortable , il faut, camarades, opposer cet exemple.*

*"A dix-huit ans le camarade Van, admissible à Normale Supérieure Science, n'a pu supporter l'idée de l'isolement du camarade Trotsky qui se trouvait alors à Prinkipo, et, dédaignant d'assurer son propre avenir, lui a offert spontanément ses services. Il l'a suivi partout dans son exil, il a passé par les mains de presque toutes les polices de Europe. A l'heure actuelle, très pauvre, puisque Trotsky n'est en mesure d'assurer à ses secrétaires que la nourriture et le lit, il continue à vivre sans pouvoir disposer le moins du monde de lui-même, privé même du sourire de son enfant.*

*"C'est de la meilleure grâce du monde qu'il prend sur lui une tâche écrasante: dix à douze heures de travail et comme doit être assurée sans cesse la surveillance de la maison, quatre heures de garde de nuit.*

*"Le camarade Van est un de ces révolutionaires de la tête aux pieds comme les veut Trotsky. Dans la détente de ce soir-là au restaurant, tandis qu'évoluaient autour de nous les servantes aux blouses ornées de broderies éclatantes à la mode de Tehuantepec, il subissait sans se départir de son beau sourire mon flot de*

*questions. Pour nous consoler de tant d'autres, il était vraiment l'homme, tel que je l'entends, l'ami dans toute l'acception du mot."*

163 Diego Rivera knew French (he had lived in France for more than ten years) but his French was probably not letter perfect. By 1938, Van Heijenoort spoke Spanish well and, most likely, wrote it perfectly.

164 "friendly ... coup d'état": Jean van Heijenoort. *With Trotsky in Exile.* Cambridge: Harvard University Press, 1978, p. 137.

165 "The two men never ... spoke": In an interview in *The New York Times,* April 15, 1939, Rivera said that the "incident" between Trotsky and him was not a quarrel but rather a "lamentable misunderstanding" which, having gone too far, became irreparable and caused him to break relations with a great man for whom he always had and would continue to have the greatest admiration and respect.

### Chapter Nine: Two Women

168 "Ah, Frida.": JvH interview with ABF, 1984.

172 The quarrel between Trotsky and Rivera: Jean van Heijenoort. *With Trotsky in Exile.* Cambridge: Harvard University Press, 1978, p. 138. Diego Rivera. "Rivera Still Admires Trotsky," *The New York Times,* April 15, 1939.

172 Rent dispute: Hayden Herrera. *Frida.* New York: Harper & Row, 1983, p. 473.

173 Trotsky's letter to Frida: Hayden Herrera. *Frida.* New York, Harper and Row, 1983, p. 248.

174 JvH interview with ABF, 1984.

174 Loretta Guyer: Interview with ABF, 1983.

175 "He looked like a movie star": Loretta Guyer, interview with ABF, 1983. Stefan Bauer-Mengelberg, who met van Heijenoort in 1947, also reported that the resemblance to Jean Marais was the first thing that struck him.

176 "sad period in Frida's life": Hayden Herrera. *Frida.* New York: Harper and Row, 1983, pp. 272–286.

178 "Bunny was thrilled": Loretta Guyer: interview with ABF, 1983.

179 "Nowhere in the book is there a mention of Loretta Guyer.": Although van Heijenoort rarely focused on his own personal life in *With Trotsky in Exile,* he did devote some space to Gabrielle Brausch and his first marriage. Perhaps he did not write of his second wife and their marriage as a matter of discretion since Bunny was alive when he wrote the book.

### Chapter Ten: Out of the Shadow?

186 "Famous and glamorous": Joel Carmichael, interview with ABF, 1983.

187 "Forty-second Street": Interview with JvH, 1983

188 "Lists of books": Sidney Hook, interview with ABF, 1989.

189 Ann Vincent is closer to his true name than any other, and, is almost anagrammable to Van Innocent. Could vH have been making a pun?

189 *The Fourth International*: Published by the National Committee of the Socialist Workers Party, New York, 1941–45. See works by Jean van Heijenoort in the Bibliography for a list of articles written by JvH under pseudonyms.

192 "Darkness set in.": Jean van Heijenoort. *With Trotsky in Exile*, Cambridge: Harvard University Press, 1978, p. 147.

194 "miraculous escape": Esteban (Sieva) Volkov, interview with ABF, 1988.

195 Letter from Trotsky in Coyoacán to JvH in New York, Aug. 2, 1940, Trotsky Archive, Houghton Library, Harvard. This letter was written only a few weeks before the actual assassination, but van Heijenoort's offer must have come a few months earlier. It seems likely that Trotsky would have responded to that first offer in some way earlier than the date of this letter and that what is here recorded is a follow-up response.

197 The onset of World War II: JvH *fils*, interview with ABF, 1983.

198 "There was no more milk for breakfast": JvH *fils*, interview with ABF, 1983. Recalling his earlier reference to ovaltine for breakfast in Coyoacán, he added, "It always seems to be breakfast time in my memory."

198 "My father spoke no French": JvH *fils*, ibid.

199 "Rarely...visit wife's family": Loretta Guyer, interview with ABF, 1983.

200 "Aside from good hands": JvH *fils*, interview with ABF, 1983.

203 "And the food!": Interview with JvH, 1983. JvH told the prison story with childish glee, taking perverse pleasure in spelling out the details of the foul prison fare. By coincidence he told the story during the course of a good meal in a wholesome California restaurant. When the waitress brought the dessert tray, Van, who knew just how far to go when he abandoned his usually careful diet, said, "Nothing chocolate," and chose apple pie with no cream. His fondness for desserts and even more, for the idea of them ("No meal is complete without something sweet at the end,") harked back to the deprivations of his youth.

204 "I guess they called Washington": Interview with JvH, 1983. Van Heijenoort did not mention his wife's involvement in his release from prison; however Bunny recalled (1990) that *she* contacted an Arizona lawyer who facilitated Van's release. This implies that Van was in touch with Bunny while he was in prison. Also in Bunny's version of the encounter with the police chief, when the chief tipped his hat and said good evening, Van is alleged to have spit on the ground.

## Chapter Eleven: Another Life

208 Jean van Heijenoort. *With Trotsky in Exile.* Cambridge: Harvard University Press, 1978, p. 149.

210 Professor Stoker: James Stoker telephone interview with ABF, 1983.

210 "Van was a loner": Joseph Keller, interview with ABF, 1984.

211 "left field ... he seemed ageless": Peter Lax, interview with ABF, 1983. Lax was fourteen years younger than JvH.

211 "efforts to improve accent": JvH *fils*, letter to ABF, 1990.

212 "*extremely* quiet": Louis Nirenberg, telephone interview with ABF, 1985.

212 "Van was a witness": Stefan Bauer-Mengelberg, Pierre Broué, interviews 1989.

213 "wounded bird": Sidney Morgenbesser, interview with ABF, 1986.

213 "cold, remote": Sidney Hook, Memorial for JvH at Stanford,1986; interview with ABF, 1988.

213 "bright young man": Stefan Bauer-Mengelberg, interviews with ABF, 1989.

215 "weak, shallow and fifty years out of date": At about the same time as he wrote his article for *Partisan Review*, JvH wrote an essay entitled "Friedrich Engels and Mathematics" (1948, see Appendix) in which he points out Engels' errors and weaknesses and ignorance in the subject. The essay was not published until 1985 when it was included in his *Selected Essays* published by Bibliopolis. In the introduction to the volume he writes: "Were I to write on this topic now, I would do it quite differently, but I let this old article of mine stand as it is."

216 "best literary magazine": *The New York Times Book Review*, 17 February, 1974.

216 Jean Vannier (Jean van Heijenoort), "A Century's Balance Sheet", *Partisan Review*, March, 1948.

217 Alan Wald. *The New York Intellectuals*, Chapel Hill University of North Carolina, 1987.

218 "He had some discussions": Loretta Guyer, interview with ABF, 1983.

218 *Partisan Review*: Marlene Kadar, *Cultural Politics in the 1930s: Partisan Review, the Surrealists and Leon Trotsky*, Ph.D. thesis, University of Alberta, 1983.

222 "no chit-chat": JvH *fils* and Loretta Guyer, interview, with ABF, 1983.

222 "best day": JvH *fils*, letter to ABF, 1990.

224 "Wonderful?": Stefan Bauer-Mengelberg, interview with ABF, 1991.

## Chapter Twelve: The House That Jean Built

228 "zee deportation list": Stefan Bauer-Mengelberg, interview with ABF, 1988.

234 "Military service": JvH *fils*, letter to ABF, 1990.

235 "Americans in France": Loretta Guyer, interview with ABF, 1983.

237 "Gabrielle Brausch": JvH *fils*, interview with ABF, 1987.

237 "Brausch and Segal": Pierre Broué, interview with ABF, 1977. Also: *Dictionnaire Biographique du Mouvement Ouvrier Français, Vol. 20*, p. 227, Paris: Les Éditions Ouvrières.

239 "This is Jean's wife": Loretta Guyer, interview with ABF, 1983.

239 "I thought of myself as American": Jvh *fils*, interview with ABF, 1983.

240 "marriage to Bunny": Loretta Guyer, interview with ABF, 1983.

241 "Jeannot did readjust": Jvh *fils*, interview with ABF, 1983.

242 "relations with Bunny": Loretta Guyer, interview with ABF, 1983.

243 "on again, off again torment": Loretta Guyer, interview with ABF, 1983.

245 "Laure van Heijenoort's reflections": Laure van Heijenoort, interviews with ABF, 1983, 1986, 1990.

246 "The grand sum of $1000.": JvH gave that figure. JvH *fils* remembers the sum as $1500. Others have reported figures as low as $500. If one imagines a figure of $100 an acre, the figure $1300 seems reasonable. Considering what happened to land and housing prices subsequently, any of these figures sounds like the wildest fantasy.

246 *"Il n'y a rien"*: Joel Carmichael, interview with ABF, 1983.

250 "twice-told tale": Stefan Bauer-Mengelberg, interview with ABF, 1988.

251 "madly in love": Anne B., interview with ABF, 1985. At Ms. B.'s request I have not used her full name.

254 "annulment": Benjamin and Jean Gollay, interview with ABF, 1983; Anne B., interview, 1985; Stefan Bauer-Mengelberg interview with ABF, 1988.

255 Laure van Heijenoort, interviews with and letters to ABF, 1987, and JvH *fils*, interviews with ABF, 1983–1990.

256 JvH "expressed regrets" about the house in interviews with ABF, 1983 and 1985.

257 "When we drove up": Joseph Stillman and Minna Kotkin, interviews with ABF, 1983.

## Chapter Thirteen: A Passion for Exactitude

260 *"Source Book"*: Jean van Heijenoort, editor. *From Frege to Godel: A Source Book in Mathematical Logic.* Cambridge, MA: Harvard University Press, 1967.

260 "lifetime of labor": Judson Webb, interview with ABF, 1986.

261 *Summaries of talks presented at the Summer Institute for Symbolic Logic, Cornell University, 1957.* Princeton, NJ: Institute for Defense Analyses, 1960.

264 "yearning for objectivity": Isaac Levi, interview with ABF, 1989.

264 "US immigration": Benjamin Gollay, interview with ABF, June, 1983.

265 "Paulette was surprised": Paulette Corpion, interview with ABF, 1987.

268 "good friend": Beverly Woodward, interview with ABF, 1986; Harvard Memorial for JvH, April, 1986.

269 "acquisition of documents": Douglas Bryant, interviews with ABF, 1986–88.

271 "Columbia University": Sidney Morgenbesser, interviews with ABF, 1986–87.

271 "*his* exposition": Fred Flashman, interview with ABF, 1988.

273 "colleague and ally": Jacques Cohen, interview with ABF, 1989.

274 "lost interest": Laure van Heijenoort, interview with ABF, 1989.

276 "over a period of six years": Stefan Bauer-Mengelberg, interviews with ABF, 1983–1991. (To Van's great regret, Mengelberg never completed a Ph.D. in mathematics or philosophy. Well after the *Source Book* was finished, Stefan, at age fifty, went to law school and in 1979 became a Doctor of Jurisprudence.)

280 "country retreat": Sidney Morgenbesser, interview with ABF, 1986.

281 "contribution": Willard Van Orman Quine, Harvard Memorial for JvH, April, 1986.

281 "Our library": Judson Webb, interview with ABF, 1986.

282 "rumors": Judson Webb, interview with ABF, 1986.

282 *Kurt Gödel. Collected Works.* Editors: S. Feferman, J.W. Dawson, Jr. S.C. Kleene, G.H. Moore, R.M. Solovay, J. van Heijenoort. New York: Oxford University Press. Vol. I, 1986, Vol. II, 1990.

## Chapter Fourteen: The Archivist in Spite of Himself

284 "History is bunk": JvH, interview with ABF, 1984.

285 "Trotsky's papers": Pierre Broué, interview with ABF, 1984.

289 "archives...burglarized": Jean van Heijenoort. *With Trotsky in Exile.* Cambridge, MA: Harvard University Press, 1978, p. 102.

289 Rudolf Klement's death: Ibid. p. 129.

293 "Van's advice and help": Douglas Bryant, interviews and correspondence with ABF, 1986–1991.

294 Douglas Bryant: Harvard Memorial for JvH, April, 1986.

295 "an audience": *Harvard Magazine,* February, 1990.

295 Jean van Heijenoort. "The History of Trotsky's Papers." *Harvard Library Bulletin,* July, 1980.

296 "all hell broke loose": Rodney Dennis, interview with ABF, 1986.

297 "speech about mistakes": Rodney Dennis, interview with ABF, 1986.

298 "In an appendix": Jean van Heijenoort. *With Trotsky in Exile.* Cambridge, MA: Harvard University Press, 1978, pp. 151–160.

299 Warren Goldfarb: Letter to ABF, April 29, 1989.

299 Pierre Broué: Interviews and correspondence with ABF, 1984–1989.

302 Marlene Kadar: Interviews and correspondence with ABF, 1983 and 1987.

303 Marlene Kadar: *Cultural Politics in the 1930s: Partisan Review, the Surrealists, and Leon Trotsky,* Ph.D. Thesis, University of Alberta, 1983.

304 "Hoover collections": Elena Danielson, interviews with ABF, 1984–1987.

305 Boris I. Nicolaevsky collection, Hoover Institution Archives, Stanford.

305 "big hole punched in the archives": "A big hole was punched in the archives in 1933. We went from Turkey to France in July 1933. The French government had granted a visa to Trotsky, with no explicit restriction. Nevertheless, the attitude of the French police presented an element of uncertainty. When the crates containing the archives were opened, Leon Sedov, Trotsky's son, took out what I would call the sensitive folders, those of the leaders of the French and German Trotskyist groups ... and put them in what he considered to be a safe place. These folders never found their way back into the archives. Sedov also took out his own folder. He had left Prinkipo for Berlin in February 1931. Between February 1931 and July 1933 Trotsky had an extensive correspondence with him; two or three times a week, he would dictate a letter to his son to the Russian typist. These letters were very rich in content. The Sedov folder never came back to the archives, and to this day I do not know where it is.

"In the night between 6 and 7 November 1937, a number of packages containing papers were stolen from a house where they had been stored, in the rue Michelet in Paris. The report of the French police stated that the door had been cut and the whole operation conducted with professional care. A gang of professional burglars had been sent from Moscow. Through Mark Zborowski, a Stalinist secret agent who had gained Sedov's confidence, Stalin's police knew in advance what there was in the house on the rue Michelet ... So perhaps some of the folders taken out of the archives by Sedov were among the papers stolen from the rue Michelet." Jean van Heijenoort: "The History of Trotsky's Papers." *Harvard Library Bulletin*, July 1980.

306 "Nico stole them": Pierre Broué interview with ABF, 1989.

309 Letters between Trotsky and Liova: The following are examples of what is in the Nicolaevsky collection of letters from Trotsky to Liova: In late 1932 and early 1933, Trotsky wrote to Liova about Zina's (T's daughter) deep depression and insisted that Zina must return to Russia to be treated (Zina wanted to remain in Germany). After Zina's suicide, Trotsky wrote that they must not hide the facts about her death, and blames the suicide on Stalin and his cohorts. In 1935, Trotsky gave Liova, who was a student at the Sorbonne, advice about exams, urged him to quit his job, and devote himself full time to his studies. There are, also, letters chiding his son for not writing: "You promised mother to write long ago. She is very worried."

## Chapter Fifteen: The Door to Life

314 "big love story": Anne-Marie Zamora Laurent, interview with ABF, 1984.

315 "purchase of Natalia Sedova's papers": Douglas Bryant, interview with ABF, 1986.

315 "As political comrades": Anne-Marie Zamora Laurent, interview with ABF, 1984.

316 "A different person": Estaban (Vsievolod) Volkov, interviews with ABF, 1988, 1990.

317 "When I saw him": Anne-Marie Zamora, interviews with ABF, 1984.

318 "Very charming": Benjamin and Jean Gollay, interview with ABF, 1983.
319 "When he was with her": Laure van Heijenoort, interview with ABF, 1983.
320 "Like her father": Anne-Marie Zamora, interview with ABF, 1984.
320 "sweet little girl": Estaban Volkov, interview with ABF, 1990.
322 "Someone very pale": Daniela Montelatici Prawitz, interview with ABF, 1989.
323 "All mulberry velvet": Jean Gollay, interview with ABF, 1983.
325 "Loss of popular culture": Felipé Bracho, interview with ABF, 1990.
326 "swimming pool": Stefan Bauer-Mengelberg recalled Van's carefully considered choice of a pool cover as one example of his deep interest in the house and garden.
328 "Beautiful place": Laure van Heijenoort, letter to ABF, March 15, 1987.
329 "if the *gringos* asked": Ezequiel Valdés, interview with ABF, 1990.
329 "I had no idea who they were": Salvador Núñez Trasloheros (Padre Chavo), interview with ABF, 1990.
332 "absolutely right place to live": Anne-Marie Zamora Laurent, interview with ABF, 1984.

## Chapter Sixteen: The End

337 "*café au lait*": Patricia Griscom, interview with ABF, 1987.
338 "What kind of fool": Pierre Broué, interview with ABF, 1984.
339 "*Je traverse une période*": letter from JvH to Anne Fagot-Largeault, 25 March, 1983.
340 "Don't you think she looks like me?": Patricia Griscom, interview with ABF, 1987.
342 "She didn't smile": Patricia Griscom, interview with ABF, 1987.
343 "married very soon": Anne-Marie Zamora, interviews with ABF, November 1984.
344 "afraid of growing old": Françoise Ville, interview with ABF, 1990.
344 "They could never": Anne Fagot-Largeault, interview with ABF, 1990.
345 "appreciation of Van's expertise": Olivia Gall, letter to ABF, February, 1987.
346 "another side of Van": Olivia Gall, letter to ABF, February, 1987.
347 "something clicked": Patricia Griscom, interview with ABF, 1987.
348 "He was in very good spirits": Anne Fagot-Largeault, interview with ABF, 1987.
350 "enormous pull": Patricia Griscom, interview with ABF, 1987.
351 "a great man": Giovanna Corsi, interview with ABF, 1986.
353 "mystified": Laure van Heijenoort, interview with ABF, 1986.

355 "crise conjugal": Lea Beaussier, letter to ABF, March, 1987.

355 "sign of trouble": Paulette Corpion, interview with ABF, 1987.

355 "Anne-Marie lamented": Laure van Heijenoort, interview with ABF, 1987.

358 "me too": Pierre Broué, interview with ABF, 1989.

359 Olivia Gall, interviews and correspondence with ABF, 1986-1990.

360 "It was a disaster": Philippe de Rouilhan, interviews with ABF, 1987 and 1990.

360 "He came to see me": Françoise Ville, interview with ABF, 1990.

360 "haggard and exhausted": Patricia Griscom, interview with ABF, 1990.

361 "it was over": Giovanna Corsi, interview with ABF, 1986.

363 Peter Katel, *Mexico City News,* April 13, 1986.

363 "homicide–suicide": Police Report, Federal District, Mexico City, March 30, 1986.

365 "more than casually aware": Peter Katel, letter to ABF, April 1987. By yet another coincidence, Peter Katel's mother, Helen, had met Jean van Heijenoort in a bookstore in France in 1933. She and a Stalinist were engaged in a loud argument which van Heijenoort overheard. He passed her a note saying he would like to talk to her outside, but when he suggested that she join his group of Trotsky supporters, she told him she already belonged. Years later, Helen and Van met again, in New York. He was married to Bunny and she was married to Jacques Katel, a French journalist at the United Nations. The two couples became close friends and their children played together. Thus, from his parents as well as his own personal contact, Peter knew of Van's history and perigrinations. Although they never met in Mexico, Peter, who lived in New Mexico in the late 1970s, saw Van several times when he came to visit Laure, who also lived there.

365 "Desolated by their loss": Estaban Volkov, interview with ABF, 1990.

## Notes for the Afterword:

367 "But in the following months": The dates of the memorial gatherings were: Stanford, April 18, 1986; Harvard, April 28, 1986; Institut Henri Poincaré, May 14, 1986.

367 "an imbalance that troubled some of the logicians": Pascal Engel, letter to ABF, October 1991. "It was a curious occasion in which people from vH's two lives were present: former trotskyists and logicians and philosophers. Both groups seems to ask what the others were doing there. Only JvH could connect them but it is my impression that he never discussed the topics of one life with the protaganists of the other. It is my feeling that these two lives or this double connection is in part why JvH was, when one approached him, so distant and mysterious. I always had the feeling that when we were discussing logic that he could 'close the curtain' (as you say) and think about Trotsky and his life, and the reverse. I interpret his reserve and silences as a sort of jump from one life scene to another."

368 "a paean to a fallen comrade": Ruth Barcan Marcus, letter to ABF, May, 1986.

368 "But the women ... knew better": perhaps the sharpest comment came from Anne B. who asked, "Was he responsible in some way? ... He did have a history of driving his women crazy although not always literally." letter to ABF, May 23, 1986.

369 "None of these problems ...": I think Van was happy to shove his problems under the rug that night.

# Bibliography

## WORKS BY JEAN VAN HEIJENOORT

### Books and Monographs


1949 *On Locally Convex Surfaces.* Ph.D. Thesis, New York University.

1967 Editor of *From Frege to Gödel: A Source Book in Mathematical Logic, 1879–1931.* Cambridge, MA: Harvard University Press.

1968 Editor of *Jacques Herbrand. Écrits Logiques.* Paris: Presses Universitaires de France.

1970 Editor of *Frege and Gödel: Two Fundamental Texts in Mathematical Logic.* Cambridge, MA: Harvard University Press.

1976 *El Desarrollo de la Teoría de la Cuantificación.* Instituto de Investigaciones Filosóficas, Universidad Nacional Autónoma de México.

1978 *With Trotsky in Exile: From Prinkipo to Coyoacán.* Cambridge, MA: Harvard University Press.

1979 *Introduction á la Sémantique des Logiques Non-classiques.* Paris: Collection de l'École Normale Supérieure de Jeunes Filles.

1980 Editor of *Léon et Natalia Trotsky: Correspondance 1933–1938.* Paris: Gallimard.

1984 *Hacia una Explicación de las Entitades Lógicas.* Colegio de Filosofia, Universidad Nacional Autónoma de México.

1985 *Selected Essays.* Naples: Bibliopolis.

1986 Co-editor (with Solomon Feferman, John W. Dawson, Jr., Stephen C. Kleene, Gregory H. Moore, Robert M. Solovay) of *Kurt Gödel. Collected Works, Vol. I, Publications 1929–1936.* New York: Oxford University Press.

1990 Co-editor (with S. Feferman, *et al*) of *Kurt Gödel. Collected Works, Vol. II, Publications 1938–1974.* New York: Oxford University Press.

Articles


1948 "A Century's Balance Sheet." (pseud. Jean Vannier). New York: *Partisan Review*, 1948 *Vol. 15*, pp. 288–296.

1948a "Friedrich Engels and Mathematics." (Previously unpublished ms.) *Selected Essays*, 1985, pp. 123–151.

1952 "On Locally Convex Manifolds." *Communications on Pure and Applied Mathematics, Vol. 5*, pp. 223–242.

1963 "Gödel's Theorem." *Encyclopedia of Philosophy, Vol. 3*, New York: Macmillan, pp. 348–357.

1963a "Logical Paradoxes." *Encyclopedia of Philosophy, Vol. 5*, New York: Macmillan, pp. 45–51.

1967a "Logic as Calculus and Logic as Language." *Synthese, Vol. 17*, pp. 324–330.

1974 "Subject and Predicate in Western Logic." *Philosophy East and West, Vol. 24*, pp. 253–268.

1976a "Set-theoretic Semantics." *Logic Colloquium,'76*, Amsterdam: North-Holland, pp. 183–190.

1977 "Sense in Frege." *Journal of Philosophical Logic, Vol. 6*, pp. 93–102.

1977a "Frege on Sense Identity." *Journal of Philosophical Logic, Vol. 6*, pp .103–108.

1979a "Ostension and Vagueness." (Previously unpublished ms.) *Selected Essays*, 1985, pp. 71–73.

1979b "Absolutism and Relativism in Logic." (Previously unpublished ms.) *Selected Essays*, 1985, pp. 75–83.

1980a "The History of Trotsky's Papers." Cambridge, MA: *Harvard Library Bulletin.*

1982 "L'Oeuvre Logique de Jacques Herbrand et Son Contexte Historique." *Logic Colloquium, '81, Proceedings of the Herbrand Symposium*, Amsterdam: North Holland, pp. 57–85.

1985a "Frege and Vagueness." *Frege Synthesized: Studies of the Philosophical and Foundational Work of Gottlob Frege.* Dordrecht: Reidel, pp. 31–45.

1987 "Système et Métasystème Chez Russell." *Logic Colloquium, '85*, Amsterdam: North-Holland, pp. 111–122.


**Note:** All of the above papers except those dated 1948, 1980a, 1982, and 1987 are reproduced in *Selected Essays*, 1985. In addition, that volume contains a revised English translation of the 1982 paper and four other previously unpublished notes. For reviews and abstracts, see the book by Irving H. Anellis referred to on p. 389 above.

Articles published in *The Fourth International,* New York: Socialist Workers Party, 1940–1945 (written by Jean van Heijenoort under several pseudonyms).

Jarvis Gerland:
"The Algebra of Revolution." May, 1940.
"'Science'—Burnham's Style." June, 1940.
Vladimir Ivlev:
"The Crises in Soviet Industry." May, 1941
Marc Loris:
"Perspectives for Europe." July, 1941.
Karl Mayer:
"Lev Davidovich." August, 1941.
Marc Loris:
"France: First Signs of the Storm." October, 1941.
"Centrism and Its Future." November, 1941.
"Capitalist Economy in War." December, 1941.
"Europe Under the Iron Heel." February, 1942.
"The Riom Trial: The Truth About French Democracy." April, 1942.
"The I.L.P. [Independent Labor Party]-Words and Reality." June, 1942.
"On Some Critics of Trotsky." August, 1942.
"The National Question in Europe." September, 1942.
"Revolutionary Tasks Under the Nazi Boot." November, 1942.
"North Africa: A Lesson in Democracy." December, 1942.
"Darlan and the Liberals." January, 1943.
"The Political Misadventures of the French Bourgeoisie." March, 1943.
"The Giraud—deGaulle Dispute." July, 1943.
"Lebanon's Fight For Independence." January, 1944.
Ann Vincent:
"How the Fourth International Was Conceived." August, 1944.
Daniel Logan
"Whither France?" September, 1944.
"Discussion Article on Europe." January, 1945.
"On the European Situation." February, 1945

## Selected Bibliography

Ali, Tariq and Phil Evans. *Trotsky for Beginners.* New York: Anchor Press/Doubleday, 1982.

Barrett, William. *The Truants.* New York: Anchor Press/Doubleday, 1982.

Chiaramonte, Nicolas. *The Paradox of History.* Philadelphia: University of Pennsylvania Press, 1985.

Deutscher, Isaac. *The Prophet Armed: Trotsky, 1879–1921,* 1954. *The Prophet Unarmed: Trotsky, 1921–1929,* 1959. *The Prophet Outcast: Trotsky, 1929–1940,* 1963. New York: Oxford Press.

Flanner, Janet. *Paris Was Yesterday 1925–1939.* New York: Viking Press, 1972.

King, David. *Trotsky: A Photographic Biography.* Oxford, UK: Basil Blackwell. 1986.

Herrera, Hayden. *Frida: A Biography of Frida Kahlo.* New York: Harper and Row, 1983.

Howe, Irving. *Leon Trotsky.* New York: Viking Press, 1978.

Lottman, Herbert R. *The Left Bank.* Boston: Houghton Mifflin Company, 1982.

Pascal, Blaise. *Pensées.* Paris: Larousse, 1965.

Serge, Victor and Natalia Sedova Trotsky. *The Life and Death of Leon Trotsky.* New York: Basic Books, 1973.

Smith, Irving H., editor. *Trotsky: Great Lives Observed.* Englewood Cliffs, NJ: Prentice Hall, Inc. 1973.

Thirion, André. *Révolutionnaires Sans Révolution.* Paris: Editions Robert Laffont, 1972.

Trotsky, Leon. *My Life.* New York: Pathfinder Press, 1970.

Trotsky, Leon. *Trotsky's Diary in Exile: 1935* (translated by Elena Zarudnaya). Cambridge, MA: Harvard University Press, 1958.

Wald, Alan M. *The New York Intellectuals.* Chapel Hill, NC: University of North Carolina Press, 1987.

Wilson, Edmund. *To the Finland Station.* New York: Harcourt Brace Jovanovich, 1940.

Wolfe, Bertram D. *Three Who Made a Revolution.* New York: Stein and Day, 1984.

Wolfenstein, E. Victor. *The Revolutionary Personality: Lenin, Trotsky, Gandhi.* Princeton, NJ: Princeton University Press, 1967.

Zweig, Stefan *The World of Yesterday.* New York: Viking Press, 1943; Univ. of Nebraska Press, 1964.

# Name Index